D1212778

Sustainable Development Goals Series

World leaders adopted Sustainable Development Goals (SDGs) as part of the 2030 Agenda for Sustainable Development. Providing in-depth knowledge, this series fosters comprehensive research on the global targets to end poverty, fight inequality and injustice and tackle climate change.

Sustainability of Future Earth is currently a major concern for the global community ans has been a central theme for a number of major global initiatives viz. Health and Well-being in Changing Urban Environment, Sendai Framework for Disaster Risk Reduction 2015–2030, COP21, Habitat III and Future Earth Initiative. Perceiving the dire need for Sustainable Development, the United Nations and world leaders formulated the SDG targets as a comprehensive framework based on the success of the Millennium Development Goals (MDGs). The goals call for action by all countries, poor, rich and middle-income, to promote prosperity while protecting the planet earth and its life support system. For sustainability to be achieved, it is important to have inputs from all sectors, societies and stakeholders. Therefore, this series on the Sustainable Development Goals aims to provide a comprehensive platform to the scientific, teaching and research communities working on various global issues in the field of geography, earth sciences, environmental science, social sciences and human geosciences, in order to contribute knowledge towards the current 17 Sustainable Development Goals.

Volumes in the Series are organized by the relevant goal, and guided by an expert international panel of advisors. Contributions are welcome from scientists, policy makers and researchers working in the field of any of the following goals:

No poverty
Zero Hunger
Good Health and Well-Being
Quality Education
Gender Equality
Clean Water and Sanitation
Affordable and Clean Energy
Decent Work and Economic Growth
Industry, Innovation and Infrastructure
Reduced Inequalities
Sustainable Cities and Communities
Responsible Consumption and Production
Climate Action
Life Below Water
Life on Land
Peace, Justice and Strong Institutions
Partnerships for the Goals

The theory, techniques and methods applied in the contributions will be benchmarks and guide researchers on the knowledge and understanding needed for future generations. The series welcomes case studies and good practices from diverse regions, and enhances the understanding at local and regional levels in order to contribute towards global sustainability.

More information about this series at http://www.springer.com/series/15486

Haris Alibašić

Sustainability and Resilience Planning for Local Governments

The Quadruple Bottom Line Strategy

Haris Alibašić
University of West Florida
Pensacola, FL, USA

ISSN 2523-3084 ISSN 2523-3092 (electronic)
Sustainable Development Goals Series
ISBN 978-3-319-72567-3 ISBN 978-3-319-72568-0 (eBook)
https://doi.org/10.1007/978-3-319-72568-0

Library of Congress Control Number: 2018935915

Printed on acid-free paper

This Springer imprint is published by the registered company Springer International Publishing AG part of Springer Nature.
The registered company address is: Gewerbestrasse 11, 6330 Cham, Switzerland

Dedicated to my wonderful wife Katherine; our kids Jakub, Lamija, Imana, and Harun; and my parents Emira and Dževad for their unconditional love and support. In memory of my late grandparents Fatima and Jakub Alibašić, Fatima and Rahman Ibraković who taught me to love the Earth and respect the nature and the environment.

"These city lights.
They shine as silver and gold Dug from the night"
The Unforgettable Fire, U2
Written by Adam Clayton, Dave Evans, Larry Mullen,
Paul Hewson

And the Firmament has God raised high,
And God has set up the Balance
Quran: Chapter 55, Ar-Rahman (The Most Gracious), 7–8

Preface

This book is intended to provide guidance, practical models, and methods in applying sustainability and resilience planning in local governments to ensure the success of organizations and communities. The book provides an introduction to the concepts of sustainability and resilience and related planning methodologies. Practical ideas, solutions, and methods in applying sustainability and resilience in planning successful organizations and communities are offered throughout the book. As local governments work to balance the provision of critical services and enhance the quality of their services, replacing strategic plans with more long-term sustainability-driven plans has become paramount. The purpose of this book is to provide the essential planning tools for organizations to be able to develop sustainability and resilience plans. I hope public administrators in cities and local governments, educators, and managers and staff in nonprofit organizations benefit from the book. Sustainability and resilience planning enhances leadership, efficiency in organizations, and communities resilience by addressing good governance, environmental concerns, social issues, and sustainable economic growth.

The target audience of this book is a diverse group of professionals, elected officials, educators, and managers, divided into four categories:

1. Elected and appointed local government officials, local government planners, city managers, sustainability directors, budget/financial directors, and community leaders
2. University administrators, president and provost's office representatives, and sustainability officers
3. Managers and staff of private and nonprofit organizations
4. Academics and educators studying and teaching practical applications of sustainability and resilience

The book attempts to answer the following underlying assumptions and questions.

- Do sustainability and resiliency enhance and support the long-term success of organizations?
- How does a sustainability and resilience plan work? Do they have to be separate plans or can sustainability and resilience planning be incorporated into a single plan?
- What type of sustainability and resilience planning works?
- Where do administrators begin the preparation for both sustainability and resilience?
- How might cities and organizations benefit from having reliable and robust sustainability and resilience plans?
- When would organizations and communities deploy such plans and what is the purpose of the plan?
- How do organizations measure sustainability and resilience progress?
- What type of activities, programs, and policies should be considered under sustainability and resilience planning?

The recommendations in this book are based on years of experience in applied resilience and sustainability planning in local government, working as the director of Energy and Sustainability for Grand Rapids, MI, and years of researching and teaching sustainability and resilience courses at various universities. I learned from the best in the field of sustainability and resilience planning, from elected and appointed officials, staff, business owners, colleagues, family, and friends. This book draws on experiences in practical application and implementation of sustainability, and resilience-related initiatives in my role as a sustainability and energy director for the local government for over a decade.

Additionally, the book draws heavily from years of examining various aspects of sustainability and resilience, from climate change, climate preparedness, triple bottom line impact, quadruple bottom line, greenhouse gas emission strategies, climate adaptation and climate mitigation to sustainable energy and Lean practices. It also builds on years of researching the best practices around the United States. In addition to examining the emerging, practical scenarios from Grand Rapids, I selected other cities with sustainability and resilience practices in place reaffirming underlying notions in the book. For years, while serving on various committees and supporting the work of the Michigan Green Communities, Resilient Communities for America, Great Lakes and Saint Lawrence Cities Initiative, ICLEI, and other organizations, I had the chance to work with their representative and staff and learn from them. Moreover, I had a chance to study emerging practices from other local governments. Solutions and emerging practices from various communities are presented throughout the book where appropriate and where reports and results are readily available.

While sustainability and resilience are the focus of this book, the natural progression of sustainability into resilience as a result of years of cities and local governments trying to infuse climate change data into sustainability-related efforts is also reviewed. Also examined are greenhouse gas emissions mitigation, resilience, mitigation, and adaptation strategies. Varying terminologies are used to describe strategies to implement resilience and sustainability initiatives, from sustainability plan or resilience plan to climate action plan or climate mitigation and adaptation plan. While resilience, in general, is the ability of organizations to withstand pressures, recover, and continue operating despite disasters, natural or human-made, changing shifts in economic cycles, climate and extreme weather. Moreover, resilience is expressed as the strength with which cities and regions recover from catastrophes exacerbated by changing climate, global warming, rising seas, and extreme weather patterns. Planning for resilience and sustainability is interwoven into the fabric of local governments. Throughout the book, I refer to cities' sustainability and resilience efforts interchangeably. While sustainability planning varies from resilience planning, ideally, cities would strive toward using the quadruple bottom line approach to sustainability planning to transition to resilience planning, utilizing a single plan. Regardless of the terminology used to describe plans, the ultimate goal of sustainability and resilience planning is more sustainable and resilient organizations and communities.

Another feature of this book is that it provides a review of sustainability and resilience literature to provide context for sustainability and resilience planning. The unique nature of this book is that a professional working in a local government could follow the outline and step-by-step process provided to proceed with creating and then fully implementing a sustainability and resilience plan. Sustainability and resilience planning brings about cohesiveness to organizations and agencies to better manage their projects, policies, and programs. While resilience planning represents the next stage in sustainability planning and includes all the elements of sustainability, understandably organizations remain committed to sustainability plans. In such instances, it is vital that the sustainability plan be reinforced with climate adaptation, mitigation, and preparedness actions and strategies to achieve resilience.

Pensacola, FL, USA Haris Alibašić
January 29, 2018

Acknowledgments

I am grateful to all local governments staff and appointed and elected officials for their willingness to share information. In particular, I am thankful for the opportunity to have worked with so many dedicated city employees, elected and appointed officials, staff, community leaders, and friends and colleagues in Grand Rapids, MI. I am thankful to Springer's staff for their guidance and encouragement. I am forever grateful to everyone not mentioned here but who contributed in some fashion to the successful completion of this book.

Contents

About the Author

Dr. Haris Alibašić is an Assistant Professor in the Public Administration program at the University of West Florida. Dr. Alibašić brings 22 years of expertise and experience in the public sector, including working for the United Nations Mission and the Office of High Representative in Bosnia and Herzegovina and directing energy, sustainability, and legislative affairs policies and programs for Grand Rapids, the second largest city in Michigan. In Grand Rapids, he promoted sustainable policies resulting in significantly reduced energy usage and cost and spurring significant renewable energy investments.

Dr. Alibašić is a founder of Alibasic & Associates, LLC, a consulting firm offering services in resilience and sustainability planning, strategic planning, and sustainable energy policy. The firm provides expert advice and consultation to local governments, nonprofit organizations, and private sector organizations on effective resilience and sustainability planning, sustainability reporting, measuring, and energy policies and programs. It provides clients with a step-by-step analysis of the organizational and community resilience and short-term and long-term resilience planning.

Dr. Alibašić has over 14 years of experience teaching graduate and undergraduate in public policy, public administration, economic development, and sustainability courses at Grand Valley State University, Central Michigan University, and Davenport University. As an assistant professor at UWF, Dr. Alibašić teaches doctoral, graduate, and undergraduate level online and in-class courses in administrative ethics, the political economy, strategic management in administration, policy lab, public budgeting and finance, and public administration.

In 2013, Dr. Alibašić advised the Resilient Communities of America on climate resilience and went on to serve as co-chair for the energy sector of the White House Climate Preparedness and Resilience Task Force in 2014. Dr. Alibašić is a partner in the Florida League of Cities Municipal Research Program. Through the Partners in Municipal Research program, the Center for Municipal Research and Innovation serves as a link between Florida's public policy researchers and municipal governments, bridging the gap between academics and public policy makers and administrators. Dr. Alibašić has written and published extensively on the topics of ethics, integrity, corporate social responsibility, administrative evil, sustainability, climate resilience, economic development, climate change, and sustainable energy. In 2017 and 2018, he led the UWF interdisciplinary team of undergraduate and graduate students to the statewide Municipal Modernization competition organized by Florida League of Cities and held in Orlando, FL.

In March 2017, Dr. Alibašić was appointed by the City Council to the City of Pensacola's Climate Mitigation and Adaptation Task Force at the recommendation of local community members and an elected official.

Dr. Alibašić won the 2012 West Michigan Environmental Action Council (WMEAC) – the C.R. Evenson Award, and the 2011 Grand Valley State University's Sustainability Champion Award. In November 2016, he received the prestigious Sustainable Hall of Fame Merit Award from West Michigan Sustainable Business Forum. In January 2017, he was awarded an Emerging Scholar Award at the Thirteenth International Conference on Environmental, Cultural, Economic and Social Sustainability and the On Sustainability Research Network, held in Rio, Brazil. He also received a 2017 Great Lakes-Saint Lawrence Cities Initiative (GLSLCI) Certificate of Appreciation and November 28th of 2016 State of Michigan Special Tribute.

Editorial Boards for Peer-Reviewed Journals

Dr. Alibašić is a Section Editor of the Creighton Journal of Interdisciplinary Leadership. He is also an editorial review board member for Public Integrity, a double-blind peer-review journal on ethics and integrity, and a peer-reviewer for the Journal of Energy Policy and the International Journal of Climate Change: Impacts and Responses.

Degrees and Institutions

Dr. Alibašić holds a Bachelor's degree in Business Administration (BBA) in International Business and Marketing and a Master's degree in Public Administration (MPA) from Grand Valley State University (GVSU). Dr. Alibašić earned a Ph.D. in Public Policy and Administration from Walden University, where he was a recipient of the Doctoral Scholarship, Commitment to Social Change.

Dr. Alibašić is actively involved with the Bosnian American community and served as a past president of the Congress of North American Bosniaks (CNAB). He is an international expert team member of the Institute for Research of Genocide (Canada). In 2013, he received a North American Bosniaks' Special Recognition Award for outstanding contributions to the advancement of Bosniaks and Bosnia and Herzegovina.

Research

- Resilience and Sustainability Planning
- Climate Preparedness and Readiness
- Sustainable Energy Planning, Energy Efficiency, Renewable Energy
- Ethics, Integrity, Administrative Evil, Moral Inversion
- Corporate Social Responsibility

Web sites: http://alibasicandassociates.com and http://uwf.edu/ceps/departments/legal-studies-public-admin-and-sport-mgmt/our-faculty/faculty-profiles/dr-haris-alibai.html#form

Defining, Initiating, and Reviewing Sustainability and Resilience Planning

1

"Perhaps the attribute most critical to a learning organization is Experimentation, which is particularly hard for big organizations since they tend to focus on execution rather than innovation." Page 235, Exponential Organizations, Ismail Smail, Malone M.S., Geest, Y. V. (2014)

Key Questions

The first chapter of this book is aimed at answering the following underlying assumptions and inquiries:

- What is sustainability? What is resilience? What are sustainability and resilience planning and the differences and similarities between the two?
- Do sustainability and resilience enhance and support the long-term success of organizations?
- How do sustainability and resilience initiatives support organizational values, missions, and goals?
- Where does organizational leadership begin the planning process for both sustainability and resiliency?
- How might cities and organizations benefit from having sustainability and resiliency plans or a single plan that encompasses both sustainability and resilience?
- When would organizations and communities deploy a sustainability and resilience plan and what the purpose is of the plan?

Introduction

In this chapter, specific terms related to sustainability, history of sustainability and Triple Bottom Line (TBL), Quadruple Bottom Line, and greenhouse gas emissions, climate resilience, and sustainability planning are reviewed. The topics of sustainability and resilience and their impact on local governments and communities warrant persistent exploration, research, understanding, and in-depth analysis. The issue of sustainability is under an incessant review, revision, and inspection, and is often used to describe a prospective positive effect of actions undertaken by organizations and individuals. Frequently, sustainability is mistaken as the treatment of the financial impact of organizations and their corresponding activities and operations. An updated definition of sustainability with an extensive review of current sustainability literature is included. Also, a selective review of successful sustainability plans in various local governments across select communities in the United States is involved. Another critical aspect of sustainability and resilience review is the definition of the Triple Bottom Line and the Quadruple Bottom Line and the historical understanding of sustainability. The book offers reasons and examples for the expansion of the definition of the Triple Bottom Line to include the fourth pillar in understanding of sustainability and resilience planning.

The principal drivers for efficacious sustainability and resilience programs are the aptitude of communities and organizations to adapt to the changes in the environmental, societal, and

H. Alibašić, *Sustainability and Resilience Planning for Local Governments*, Sustainable Development Goals Series, https://doi.org/10.1007/978-3-319-72568-0_1

economic conditions surrounding them. Local governments use sustainability to address their constituents' needs and demands. Organizations are engaged in innovation to continue to provide quality of life services as revenues shrink. Local leaders are aware of the complex nature of urban cities and design programs in ways to address cities' sustainability needs and to enhance resiliency efforts of those cities stemming from security threats, emergencies, extreme weather, and climate change. Effective sustainability and resilience planning assists municipal leaders in addressing various internal and external pressures and apprehensions.

A Review of Sustainability, Historical Paths, and Significance

The issue of sustainability and resilience and the influence on local governments warrant exploration, research, appreciation, and in-depth analysis. The topics of sustainability and resilience are under an incessant review, revision, and inspection. Sustainability is often used to describe the combined social, economic, environmental, and governance issues within an organizational framework. Sustainability is regarded from the standpoint of its practicality and commonly misinterpreted as something as an additional burden and cost to the society. Contextually, many local governments around the world claim to use sustainability to further their operational efficiency and to address the economic, environmental, and societal impact of their actions.

As the pressures over the rising cost of energy, climate change politics, and reduced revenues intensify and effect the financial bottom line, the short- and long-term sustainability and resilience planning is seen as a solution to various. To some, the term "sustainability" conveys a certain sense of continuity that withstands the test of time. Slavin (2011) alluded to this sense of endurance in defining sustainability as "the capacity of natural systems to endure, to remain diverse and productive over time" (p. 2). The concepts of sustainability and resilience give equal weight to the environmental, social, and economic issues.

Additionally, equally important for modern organizations is the expansion of the concept of sustainability and framing it through the resilience mechanism to better understand the depth and breadth of climate change and related impacts. Resilience adds furthers the comprehension of sustainability with heightened pressures from the climate change and the effect it has on communities, often reflected through extreme weather events, infrastructure, and pressures on human resources. Resilience may best be described as an added, enhanced level of sustainability planning, by taking into account the issues of climate change.

Sustainable organizations include a commitment to pursue sustainability, a collective understanding of what sustainability constitutes, a leadership endorsement of sustainable practices, and keeping critical stakeholders engaged by maintaining the focus on the broad concept and vision that sustainability is about social, environmental, and economic health (Hitchcock and Willard 2008). More specifically, identification of shared goals and targets will further assist organizations in achieving sustainability.

Theoretical Background

Given the diverging views on sustainability, theories directly or indirectly related to sustainability and resilience are critical to the better understanding of the postulates of sustainability and resilience. Gaertner (2009) described the theory of social choice as "an analysis of the collective decision making" and contemplated aggregation of "individual preference" in the reflection of the general preference of the society (p. 1). Analyzing sustainability and related policies offers a better grasp of the measures undertaken and overall outcomes on the society or organizations. As argued by Elster and Hylland (1989), social choice theory emanated from two different problems, one of which is the "finding measure for aggregate social welfare" (p. 2). The very idea of sustainability is to find the problematic and hard to define measure, weight, and process of social welfare, through the sustainability-related lens.

Heal (1998) methodically explained the essential axioms of environmental assets, including "a treatment of the present and the future," and recognition of both "how environmental assets contribute to economic well-being" and "the constraints implied by the dynamics of environmental assets" (p. 14). This method emphasizes the environmental benefit as a substance of sustainability and does not delve into the social aspect of the Triple Bottom Line. Solow (1992) also offered a rational policy approach using economic theory to defend the notion of possible improvements to "economy about its endowment of natural resources" (p. 5). Again, the focus is on the environment, but with a clear understanding that improved environment leads to enhanced economic and societal outcomes.

In addition to social choice theory and economic theories, another theoretical framework connected to the issue of sustainability is the system theory. Von Bertalanffy (1950) introduced the idea of a general system theory and deliberated that "general system laws" apply to any system of a certain type, irrespective of the particular properties of the system, or elements involved" (p. 138). Any phenomena may be regarded as the interconnected system of different elements, whether it is sustainable energy or another process. As von Bertalanffy (1950) suggested, general system theory is "applicable to all sciences concerned with systems" (p. 139). In discussing the system theory, Patton (2002) maintained that "a system is a whole that is both greater than and different from its part" (p. 120). Such approach facilitates an explanation of the contested sustainability phenomena and the methods under which underlying elements of environmental, social, governance, and economic components function.

There are varying ideas, concepts, paradigms, and theories used to construct sustainability and sustainable development framework and corresponding economic, social, and environmental bottom line for organizations and society. Heal (1998) interpreted sustainability as "a metaphor for some the most perplexing and consequential issues facing humanity" (p. xi). A body of work from other disciplines including among others, economic, social, environmental, and political, provides a meaningful theoretical definition of sustainability.

Sustainable Development and Triple Bottom Line

The fundamental postulates of sustainability and sustainable development were first established by the United Nations' World Commission on Environment and Development. The United Nations' World Commission on Environment and Development (1987) coined the term sustainable development as the rational management of resources in the present by organizations and individuals without compromising the needs of future generations (p. 4). The Report of the World Commission on Environment and Development, *Our Common Future* from Brundtland Commission, was set up by the United Nations, which provided the original definition of sustainability.

"Development that meets the needs of the present without compromising the ability of future generations to meet their own needs," with "two key concepts," where "a process of change in which the exploitation of resources, the direction of investments, the orientation of technological development; and institutional change are all in harmony and enhance both current and future potential to meet human needs and aspirations."

In the early stages of defining sustainability, the United Nations' World Commission on Environment and Development (1987) established the sustainable development framework keeping future societal needs in mind. However, since the initial platform for sustainability was developed, a significant amount of research was invested in redefining and refining the sustainability. There are inconsistent interpretations of sustainability and its impact on organizations, communities, and society. Stubbs and Cocklin (2008) put it succinctly how "sustainability itself is a contested concept" and concluded a lack of consensus on the very definition of sustainability (p. 104).

Elkington (1997) provided pioneering and groundbreaking views on sustainability globally with his task to corporations to evaluate the environmental and social impact of their actions. Elkington's classic from 1997 under the title *Cannibals with Forks: The Triple Bottom Line of 21st Century Business* offered the first glimpse of the Triple Bottom Line definition and its potential impact on companies and organizations. Often, the three areas of influence are referred to as the Triple Bottom Line (TBL) (Elkington 1997; Savitz and Webber 2006). The Triple Bottom Line (TBL) relates to initiatives undertaken in each of the areas of economic prosperity, social equity, and environmental integrity.

From TBL to Quadruple Bottom Line (QBL)

An early concept in defining sustainability was the Triple Bottom Line approach to measuring impact from organizations on the society. Savitz and Weber (2006) viewed the Triple Bottom Line as a balanced way "that captures in numbers and words the degree to which any company is or is not creating value for its shareholders and society" (p. xiii). Elkington (1997) created the Triple Bottom Line axiom to seek of corporations, to measure, and to evaluate their social and environmental impact on the society and their environments beyond what they produce for their economic benefit. Sustainability is viewed as an opportunity for organizations and in the milieu of the necessary evolution of society. In the later writings, Elkington (2012) posited that sustainability supports better corporate governance, which in turn builds "genuinely sustainable capitalism" (p. 6).

The imperative posited through Triple Bottom Line was to challenge private sector organizations to implement goals focusing on economic prosperity, environmental protection, and social equity as a necessary objective of achieving success for corporations. Whereas Elkington's (1997) Triple Bottom Line definition focuses on the private sector, its broad application of postulates applies to the public sector organizations. However, as the sustainability evolves, its static description looking through three basic pillars of sustainability needs constant reinvention and revisiting. The proposed Quadruple Bottom Line looks at the issue of sustainability from an added perspective of focusing on governance. An expanded definition of Quadruple Bottom Line is

> Organizational capacity to embed and incorporate a set of definitive policies and programs to address economic, social, environmental, and governance aspects of sustainability, whereas governance is defined through fiscal responsibility and resilience, community engagement for effective service delivery, and transparency and accountability. Alibašić (2017)

The transition from TBL to QBL is best explained visually using the following diagram. Governance is a dynamic component necessary to the successes of sustainability and resilience. Moving to including and assessing the good governance is critical to the evolution of sustainability and resilience (Fig. 1.1).

Sustainability and Resilience of Local Governments

Early roots of the local government involvement in sustainability and a call to action on sustainability can be traced to the United Nations Conference on Environment and Development in 1992 and Report of the United Nations Conference on Environment and Development, Vol. 1, with resolutions adopted by the conference (Agenda 21) in Rio de Janeiro, Brazil. As indicated in the Agenda 21 report, "Rapidly growing cities, unless well-managed, face major environmental problems," and further "the increase in both the number and size of cities

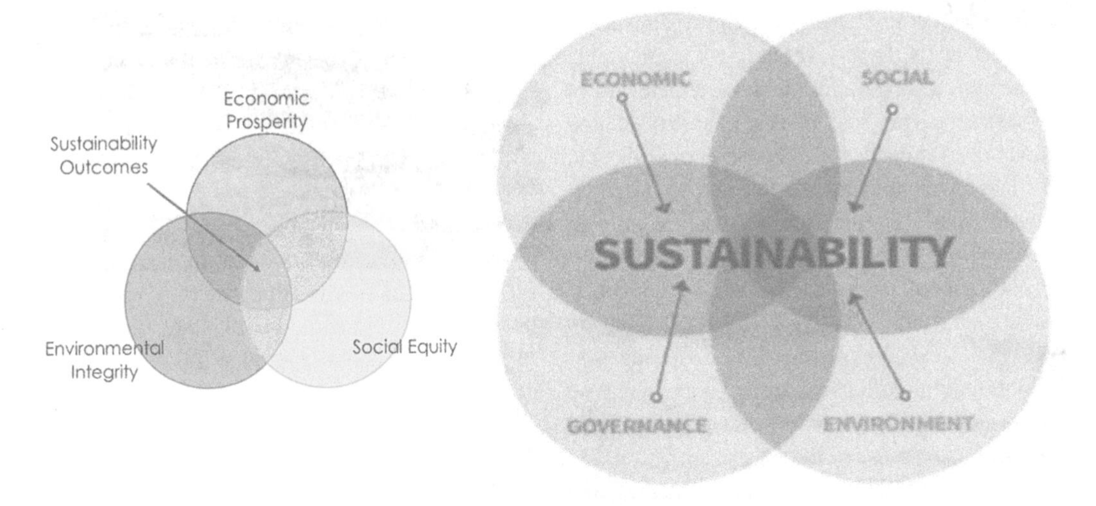

Fig. 1.1 An evolution from Triple Bottom Line to Quadruple Bottom Line

calls for greater attention to issues of local government and municipal management" (p. 1.).

One of the key objectives of Agenda 21 was "to implement policies and strategies that promote adequate levels of funding and focus on integrated human development policies, including income generation, increased local control of resources, local institution-strengthening and capacity-building and greater involvement of non-governmental organizations and local levels of government as delivery mechanisms" (p. 14). Another important aspect of Agenda 21 to the evolution of sustainability is its focus on social equity. As noted in Agenda 21, Principle 1, "Human beings are at the center of concerns for sustainable development. They are entitled to a healthy and productive life in harmony with nature" (United Nations General Assembly 1992). A balancing of social equity with a clean, healthy environment and a supporting economy create sustainability, taking into account the past, present, and future practices. Conversely, ignoring the needs (especially of our impoverished populations) would have devastating effects on the society as a whole.

More cities focus their efforts on addressing social equity in their sustainability plans:

> Social equity means that all citizens have equal access to goods and services, education, and resources that enable people to have a high-quality life. Public institutions are to provide fair, just, and equitable distribution of public services, while promoting fairness, justice and equity in the formation of public policy (City of Grand Rapids 2011, p. 17).

The goal of sustainability-related efforts is to impact not only the economic, but also the environmental, social, and to a certain extent, the governance aspects of an organization. Moreover, "Ethical implications of sustainability planning and implementing sustainability-related efforts include more just, equitable, healthy, and environmentally resilient communities with an overall positive societal outcome" (Alibašić 2018a, p. 4).

Local governments involved in sustainability are committed to protecting environmental and social resources in delivering the most efficient services. There are connections between the practical and applied implication of sustainable actions, and the impact organizations have on the community and society at large. By actively pursuing investments in renewable energy and energy efficiency, organizations positively impact their economic, social, governance, and environmental bottom line, thus affecting positive social change. Sustainability and resilience outcomes may be more feasible and attainable in cities, at local levels of government which tend to be more homogeneous in their pursuit of policies and programs regardless of political affiliation. At the very least, gaining consensus on issues at the local level appears to be viable as compared to the state and federal levels of government. The goal of sustainability-related efforts is to impact not only the economic, but also the environmental, social, and to a certain extent, the governance aspects of an organization.

With globalization and increased economic pressures on cities, collaboration is the key to success. More tangibly, elements fostering sustainable communities include cohesive land use policies, sustainability and resilience-driven partnership opportunities with the private sector and academic institutions, and solid strategies and plans in place. Support systems for these elements of sustainability include sustainable water systems, transportation systems, waste management systems, natural resources preservation, and food production (Coyle 2011). Some of the resources that are imperative to sustainable and resilient communities are energy, job opportunities, transportation, and public safety. Intangible resources include collaboration between government, population, private, and nonprofit sectors and that cooperation drives the sustainable growth. Saha (2009) pointed out to the fact that "local government sustainability initiatives have emerged in response to the growing recognition of the importance of taking local action toward global sustainability" (p. 39). Some local governments found the need to fill the void left by national governments in meeting their sustainability-related objectives in the society.

In such an environment, local governments, mainly cities, are faced with challenges that occur as a result of our civilization's dependence on energy. Coupled with the fact that "modern cities function very differently from the way cities did in the past," a very different future scenario is facing the cities and communities (Girardet 2006, p. 11). Girardet (2006) offered further understanding of a sustainable city, identifying their enormous impact on the economy and environment, through positive actions aimed at reduction of the energy demand and energy consumption.

Local governments are faced daily with crucial decisions on providing services and meeting increased demand for services while facing constant and severe budget cuts to staffing and operations. At the same time, city governments are expected to provide the same level of services without additional revenues or resources. Institutionalizing sustainability is an enormous undertaking, which requires leadership and readiness to measure, track, and report progress. Fitzgerald (2010) stated that "we have seen cities link sustainability and climate change initiatives to green job creation and even the development of whole new industries" and "how cities have inspired national policy, after a long period of federal government inaction" (p. 176). When available funding is in peril, local governments' capacity to pursue sustainability becomes an added effort, beyond required tasks to provide services. For sustainability and resilience planning to be fully embedded within organizations, it must become an integral part of the budgeting process, through active pursuit of sustainability and resilience goals and targets.

As a strategic imperative, sustainability planning has become the norm many local governments. Similarly, to Martin et al. (2012), authors Ammons et al. (2012) discussed the new normal for local governments and observed how they "confront multiple points of tension that pull local officials in different directions simultaneously and collectively influence a government's structure, scope of services, and philosophy" (p. 71S). Authors argued the long-lasting impact on local governments and their delivery of services beyond the recent economic recession and downturn in the economy (Martin et al. 2012; Ammons et al. 2012).

Sustainability in Private Sector

The pursuit of sustainability is not constrained to the public sector alone. In making the business case for sustainability for businesses, Werbach (2009) stated "the global economy, our environment, and political institutions are undergoing rapid structural change" (188). Furthermore, Waddock (2009) discussed at length "emerging corporate practices" to support companies' "path toward sustainability," including product accountability, life cycle management, and spreading the cost of emissions and benchmarking them with others (p. 303). Sustainability may be observed through a lens of long-term implementation strategy and initiatives by a given organization, with the ultimate objective of providing services and products more sustainably.

Sustainable organizations strive to have the most positive economic and societal impact while at the same time having the least negative impact on the environment. Companies also realize and find the case for sustainability and resilience in knowing the risk of failed cities, communities, infrastructure, and the society for their operations and the bottom line (Alibašić, 2018b). Corporations may be able to quickly adjust and seek opportunities for long-term solutions through ideas tied directly or indirectly to sustainability. On a large scale, problems facing companies and cities appear to be interconnected to the opportunities and woes that exist with the global economy. As the societal paradigms shift, corporations and local governments, as well as other organizations, use sustainability and resilience to support their long-term strategic goals.

Hardjono, Van Marewijk, and de Klein created The European Corporate Sustainability Framework (ECSF), who (as cited by Stubbs and Cocklin, 2008) developed "a set of models, tools, and theories—to help organizations address complex social and environmental sustainability issues" (p. 104). Hardjono et al. (2004) provided a comprehensive methodology for implementation of sustainability in organizations by capitalizing on organizational dynamics. The authors utilized symbolic interaction methodology in a systematic approach to enable companies to employ corporate sustainability and social responsibility methods. Another consistent framework modeling energy supply and demand for sustainable cities was developed by Brownsword et al. (2004) and which analyzes "both technological and socio-economic aspects of domestic and commercial energy-consumption and use the results to produce a model for urban energy-management" (p. 168). The research adds a new dimension to a methodology of evaluating sustainable energy and how it corresponds to sustainable cities. It is an insight into the role of sustainability and practical implications for organizations that organizations can then utilize when it comes to energy planning and planning for sustainability.

There are arguments that organizations, mainly corporations, use the sustainability bottom line to break ranks with the accepted views of businesses and to advocate for societal issues such as climate change and protection of the environment. Bendell and Kearnis (2005) discussed some examples of companies using their economic and business clout to pursue political agenda and to advocate for various sustainability-related issues including the climate change, impacting markets, and the rest of the society. Finally, sustainability demands a collective, collaborative effort by a broad segment of the population, including a commitment by public, nonprofit, and private sectors.

Savitz and Weber (2006) defined the Triple Bottom Line as the element of sustainable businesses. Furthermore, the authors argued, "a sustainable company manages its risks and maximizes its opportunities by identifying key nonfinancial stakeholders and engaging them in matters of mutual interest" (Savitz and Weber 2006, p. 18). Sustainability challenges corporations to assess their social and environmental impact on the society, and not just economic effect, as a collective good, and a maximized business opportunity (Galea 2004, p. 37; Adams et al. 2012, p. 17; Elkington 1997).

Starting Point(s)

Similar to strategic planning, creating a sustainability plan is not an easy and straightforward task. Developing a sustainability plan is an organic, bottom-up, linear, and engaging process. Organizations recognized for successful sustainability and resilience efforts make strategic and budgetary commitments to integrate sustainability into their goals, mission, and values. Sustainability and resilience planning becomes a practical, valued added, and applied strategy for organizations, with a benefit to the community at large. The process of embedding sustainability at all levels of local government is a long one, and it cannot be done in a vacuum. Several elements contribute to the successful application of sustainability efforts, including but not limited to:

- Internal efficiency and operational improvements using Lean principles.
- Significant policy and planning stages – sustainability plan, renewable energy goals, and green building policy.
- Community's participation and partnerships in pursuit of sustainability.
- Positive engagement of staff and key stakeholders, whether elected or appointed officials in the city hall are critically important.
- Empowering employees to champion sustainability targets.
- Measuring, tracking, and reporting results of the sustainability-related efforts, using sustainability progress reports. As part of such efforts, setting clear goals and objectives is imperative, coupled with specific targets of the plan.
- Connecting targets to the budget and fiscal plans and policies.

Each segment of creating sustainable community feeds into the next. At a minimum, the starting point should include the following list of questions.

- What are the current economic, social, environmental, and governance issues in the organization and the community?
- What projects and issues have the most impact from that organization's perspective?
- How does sustainability align with organizational goals, vision, and mission statements?
- Does the organization measure its gas emissions output, does it have a carbon footprint reduction plan, and if so when was it last updated?
- Who are the major stakeholders in the planning of sustainability?
- How would the management go about engaging stakeholders?

Resilience Planning

Climate change is the most critical issue that cities are facing. While most scientists agree on the causes and impacts of climate change, the inaction on the federal level has left many municipalities to deal with this existential threat on their own. While the perils differ from regions to regions, the continuing studies of climate change indicate explicit threats to cities around the world and in the United States (IPCC 2014; USGCRP 2014).

In 1997, the Kyoto Protocol was adopted as an international agreement to commit participating countries to reduce their greenhouse gas emission (UNFCCC 2014). The United States did not sign the deal, and more recently the United States withdrew from the already signed Paris climate accord, leaving states and communities to deal with the consequences of inaction. On August 4, 2017, in an official communication from the US Department of State, the intent to withdraw from the accord was confirmed with the United States expressing interest in continuing its participation in the ongoing and future international climate change negotiations and summits (US Department of State 2017). The United States' absence from a leadership position in the struggle against the threats and consequences of climate change is an adverse development for many communities, dealing with climate change. The primary cause of climate change is the accumulation of greenhouse gases in the atmosphere, alarmingly escalating over the last several decades due to increased consumption of fossil fuels.

As observed by USCCSP (2009),

> Consensus in the climate science community is that the global climate is changing, mostly due to humankind's increased emissions of greenhouse gases such as carbon dioxide, methane, and nitrous oxide, from burning of fossil fuels and land-use change (measurements show a 25 percent increase in the last century). Warming of the climate system is unequivocal, but the effects of climate change are highly variable across regions and difficult to predict with high confidence based on limited observations over time and space (p. 10).

It has been accurately demonstrated that human activities are the principal contributors to accelerating the increases in greenhouse gas emission. The NASA (2016) has provided a stark warning of consequences of climate change such as

> temperatures will continue to rise; frost-free season will lengthen; changes in precipitation patterns; more droughts and heat waves; hurricanes will become stronger and more intense; sea level will rise 1-4 feet by 2100; and the Arctic is likely to become ice-free.

Focus on the localization of climate change impacts a specific set of recommendations to build resilience in the community, to prepare cities, and to strengthen disaster recovery and resilience by local government. Climate resilience is attributive to an ability of absorption, healing, and adaptation to not only the adverse environmental changes but to the economic and social shocks exacerbated by the advent of climate change. It is vital for cities to possess prevention tools and resources necessary to provide immediate responses to disasters and shocks. For a resilient community, all essential supporting systems need to be adaptable, flexible, and vigilant. Therefore, the idea of resiliency holds dynamic dimension of the endurance of population and nature and allows communities to stay ahead of potential affliction.

This dynamic nature of climate change is best described in the West Michigan Environmental Action Council (2013), describing the impact of it:

> Climate change impacts each sector in isolation, but it also impacts the interaction of each to others and the function of the system as a whole. Therefore, understanding the needs of the community, major relationships between sectors, and the ability of the sectors to provide those needs in a changing climate world is key to building resilience (WMEAC 2013).

Beyond sustainability, local governments in their organizational and leadership capacity continue to deploy resilience planning, to address environmental, social, economic, and governance issues, stemming from climate change and extreme weather. Both climate adaptation and climate mitigation are included in such plans. Most successful sustainability and resilience plans cover all aspects of organizations into those programs. Recognition of climate change threats become a routine component of local planning to create more viable and resilient communities. As the Fig. 1.2 indicates, a choice is not between doing mitigation work and adapting to the change that has already occurred. Out of necessity, local governments deploy both, the climate mitigation and

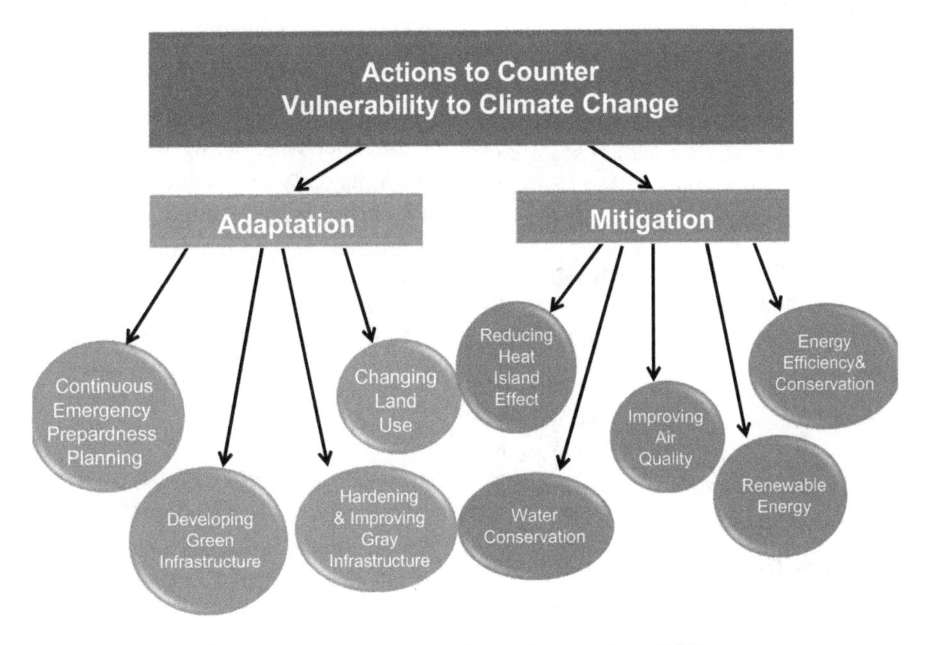

Fig. 1.2 Actions undertaken by organizations to counter climate change vulnerabilities

climate adaptation strategies. All aspects of the built environment need to be incorporated into the resilience planning. Some mitigation strategies may fall into adaptation strategies and vice versa. The preparation must include updating emergency plans to consider climate-vulnerable citizens, developing green infrastructure while hardening and expanding gray infrastructure, and anticipating events never before experienced. Local government planners should consider measures to reduce heat island effects, increase air quality, reduce water usage, and reduce energy consumption in operations.

Definition of Terms

Describing sustainability and resilience using appropriate terminology is quintessential to the embracement of those two concepts. The following are some of the most relevant terms used throughout this book.

Carbon footprint The overall amount of greenhouse gas emissions (GHG) produced by an organization (or individuals) in a given period as a consequence of the power production, heating and cooling, and transportation using fossil fuels, and dietary preferences favoring meat consumption.

Climate Change A scientific fact supported by numerous studies and research proving the climate change is occurring on a global scale and that the leading cause is human actions linked to the exploitation of traditional sources of energy.

Climate Adaptation Strategies deployed by organizations to indicate the need to adapt to changing conditions as a result of climate change.

Climate Mitigation Strategies deployed by organizations to decrease the pollutants by moving away from the use of traditional sources of energy such as coal, oil, gasoline, and natural gas in operations, including transportation.

Climate Preparedness and Readiness An overall strategy deployed by local governments and other organizations to prepare and implement strategies to combat perils stemming from the changing climate and extreme weather.

Energy Efficiency Investments made and activities undertaken to reduce energy consumption and positively impact organizations and individuals paying energy bills.

Feed in Tariff A policy explicitly developed to encourage the use and production of renewable energy, by paying an above the cost price to energy producers for renewable energy production.

Global Warming A scientific body of research showing that Earth's temperature is on the rise since the nineteenth century and is projected to increase further as a result of human activities, causing the global warming effect.

Greenhouse Gas Emissions (GHG Emissions) Gases trapping heat in the atmosphere are considered greenhouse gases (GHG) and include carbon dioxide (CO2), methane (CH4), nitrous oxide (N2O), and fluorinated gases.

Lean (in Government) A strategy employed by organizations focusing on eliminating waste in the processes and improving efficiency and efficacy of service delivery based on manufacturers' approach to streamlining operations.

Peak Oil A scientific argument made by oil and energy experts predicting that the world either reached its daily oil production capacity or will have reached soon, predicting a reversed trend in oil production and causing a future energy crisis.

Renewable Energy Energy produced from sources that appear infinite in its current form, wind, hydro, solar, and geothermal (power from underground Earth's heat).

Renewable Portfolio Standard (RPS) A policy program usually adopted by states to require energy generators/utilities to produce a certain percentage of electricity from renewable energy sources.

Resilience (Climate Resilience) An advanced approach to prepare organizations and communities for threats from climate change and planning for such risks, incorporating various economic, environmental, social, governance, and emergency preparedness strategies into sustainability and resilience planning. Such planning strategies prepare organizations and communities to withstand distresses, shocks, and disasters and to continue to function during and after the adversities.

Sustainability An ability of organizations to provide outcomes and to maintain systems comprehensively with a minimum negative social, economic, environmental, and governance effects on resources, while maximizing positive results.

Sustainable Energy An ability of organizations and society to efficiently impact their bottom line and provide positive social change through reduction in energy consumption, production of renewable energy, and efficient management of energy.

Triple Bottom Line (TBL) A concept which describes how socially responsible organizations consider the negative impact of their actions on social and environmental aspects of society and try to minimize it by using sustainability in its core mission, values, and operations. The same principle applies across sectors, nonprofit and governmental.

Quadruple Bottom Line (QBL) Organizational capacity to embed and incorporate a set of definitive policies and programs to address economic, social, environmental, and governance aspects of sustainability, whereas governance is expressed through fiscal responsibility, community engagement for effective service delivery, transparency, accountability, and more resilient organizations and communities.

Waste Minimization An ability of organizations and communities to reduce the waste designated for landfills by deploying the recycling, reusing, and repurposing strategies to reduce and minimize waste.

Resilient City Spotlight: Replacing a Strategic Plan with the Sustainability Plan

In the past, Grand Rapids had faced severe fiscal challenges, due to a lagging economy, loss of revenues, and the increasing employees costs. As a result of its transformation effort, the city can avoid severe financial issues and close a budget gap in the coming years. Despite its economic afflictions, the city was seen by many national organizations as groundbreaking for innovative efforts in sustainability and energy, a transformative organization and the community (Geary 2011; Svara et al. 2011; Knapp 2011; McCarty 2012).

For years Grand Rapids, MI, operated with its strategic plan. The discussions to introduce sustainability into city strategic planning processes started with efforts to improve internal efficiency and operational improvements using Lean processes. The first conversations about sustainability in Grand Rapids began in early 2004. Around that same time, the city started introducing "the Lean process into operations, which proved extremely valuable to the support of qualitative and quantitative outcomes in sustainability-related efforts" (Alibašić 2013).

At first, the plan was a much more of a blueprint. However, over the next years, from its first reiteration, the program grew into a full strategic plan, thus replacing the strategic plan, tossing out of the window. Through the careful process of developing "a sustainability plan, the city of Grand Rapids, Michigan has become one of the most sustainability planning-oriented communities in the United States" (Alibašić 2013). The change took time and years of planning and changing mindsets of the staff, community members, and elected officials. In his visionary and prophetic talk to the Grand Rapids' local leadership in 2006, Kent Portney stated that Grand Rapids was "poised to enter the real elite of the country – a regional and national leader for mid-sized cities." The city built upon early plans and moved into the national spotlight, focusing on sustainability. Additionally, Portney (2012) in the annual review of sustainable cities and their climate actions and plans touted Grand Rapids as a success story in the climate action and sustainability.

Goodwin (2012), in the case study for the International City/County Management Association (ICMA), concluded how Grand Rapids' approach to sustainability may not be applicable to all local governments but found it important "how the city developed targets and measures and linked them to the budget that uses Sustainability Plan to drive its budgetary decisions" (p. 3). Geary (2011) also wrote how the city's sustainability plan "connects Grand Rapids' current sustainability initiatives with the goals of different City departments and incorporates ways to measure success" (p. 7).

Over time, the city's efforts started to pay off, and the city began to gain national and international attention for its sustainability efforts. In 2010, the city received the most sustainable midsized community award from the US Chamber of Commerce and Siemens (Beeke 2010). Local media covered extensively sustainability efforts undertaken in Grand Rapids (Boomgaard 2011; Stukkie 2012; Vande Bunte 2012; Wood 2009). The city organization with its 1450 full-time employees and a budget of over $300 million is faced with challenges and constant fiscal, external, and internal pressures. At the same time, the city has significantly reduced its workforce, met lower revenues from property tax collection, and experienced budget cuts (City of Grand Rapids 2011). Moreover, despite concessions made by employees, with significant cuts to benefits, the city has been faced with a budget deficit if it failed to transform (City of Grand Rapids 2011).

In 2012, Grand Rapids received a national recognition by the US Conference of Mayors for its climate protection efforts in the large cities category, as stated by McCarty (2012) in the press release about the award that can serve as a model "for the rest of the country" (p. 25). According to Hess et al. (2010) "Grand Rapids has some of the most aggressive renewable energy goals in the nation" with an eye toward a 100% target by 2020 (p. 67). Grand Rapids committed to purchasing the 20% of its energy from renewable energy sources by 2008, and after meeting the goal, moved it to 100% by 2025 (City of Grand Rapids 2016). Commitment to renewable energy illustrated the city's leadership and the importance of cities in the discussion surrounding sound sustainable energy practices and policies. Additionally, city employees, from various departments and service groups, are directly involved in ensuring the targets, and goals from its annually amended sustainability plan are met (City of Grand Rapids 2016).

Significant Policy and Planning Stages

The city's approach to sustainability and then to planning for resilience deserves a closer historical overview. In 2002, the city adopted the new version of the master plan with substantial public engagement and input. The master plan assisted in establishing the framework for future sustainability plan, made available to the public in August of 2006, and it was intended to move Grand Rapids toward being a sustainable city with the specific policy direction to make service delivery sustainability conscious and driven. Furthermore, as reported by Alibašić (2013), "Subsequently, in 2006, the city passed a resolution establishing a sustainability policy for city-owned buildings, standardizing requirements for construction, renovation, and management, requiring the potential use of the Leadership in Energy and Environmental Design (LEED) principles, water conservation, and energy use reduction." Grand Rapids area is home to a noteworthy concentration of LEED-certified buildings with the first LEED-certified art museum, the city's Water Department's administration building, and others.

The City Remained Engaged on the International Stage In 2007, Grand Rapids was designated a United Nations University Regional Centre of Expertise (RCE) in recognition of its efforts to achieve the goals of the UN Decade of Education for Sustainable Development (DESD 2005–

2014) (UNU 2017). Another milestone was the city's publication of the Triple Bottom Line report in 2008, containing the community-wide triple bottom measurements as a benchmark for assessing progress. Equally important is the city's 2009 Energy Efficiency, and Conservation Strategy financed through Energy Efficiency and Conservation Block Grants (EECBG). The strategy contained not only specific recommendations for the city's energy conservation and efficiency improvements, but it also provided the first community and organizational greenhouse gas emissions report. It has been used ever since as the critical benchmark for measuring the city's carbon footprint (City of Grand Rapids 2009).

The renewable energy goals were intended to initially power 20% of the city's operations with renewable energy by the end of 2008 and then to power 100% of activities by 2025. The goal spurred an internal innovation revolution, as each department sought different solutions to meet the targets. Similar to Lean process and techniques, renewable energy targets drive and motivate city staff to be creative and innovative in finding sustainable measures. Despite the changes in appointed and elected officials, the city continues its commitment to the renewable energy targets and energy efficiency (Steiner 2017).

Community's Participation, Partnerships, and Pursuit of Sustainability

The ability to work in collaboration and conduct sustainability-related activities is what moves the sustainability needle in a positive direction. In 2005, together with four other organizations (Aquinas College, Grand Rapids Community College, Grand Valley State University, Grand Rapids Public Schools, and the City of Grand Rapids), the city created the Grand Rapids Area Community Sustainability Partnership as a diverse, collaborative effort to promote and share best sustainability practices in planning and operations.

From the five original members, the partnership grew to 280 members, including private, public, service, and academic organizations, "committed to work together to restore environmental integrity, improve economic prosperity, and promote social equity in the community with the goal of creating and sustaining a positive quality of life for future generations" (Community Sustainability Partnership 2017).

The city's sustainability plan, while it relates to the city as an organization with its operations and resources, also focuses on the more significant "implication of sustainability in the region and impact on the community. Collaborative efforts in the area of sustainability and resiliency in the region, energy audits, and energy efficiency improvements in neighborhood homes, increased recycling through local economic incentives, and the most recent work on resiliency report" with domestic partners are further evidence of the importance of partnerships to achieve successful sustainability-related outcomes (Alibašić 2013).

Summary

Evaluating the overall effectiveness and efficiency of sustainability and resilience planning, and the impact on social, economic, governance, and environmental bottom line enhances the appreciation of the local governments' service delivery efforts. The positive social and community impacts from sustainability and resilience planning include reduced greenhouse gas emissions resulting from the production of electricity from coal-powered plants, reduced reliance on oil imports, lessened effect on the environment, improved governance, better service delivery, ability to withstand shocks, and other societal benefits.

Cities benefit from a positive social impact as the reduced investment in energy consumption free up capital to be invested into other services provided by local governments, such as street improvements, public safety, parks and recreation to name a few. Once sustainability performance is reported, measured, and then compared to various outcomes, it assists organizations to assess the positive impact on the overall effectiveness of service delivery of local government. All aspects of societal issues can be evaluated using sustainability as a lens or a conceptual framework. The sustainability and resilience planning process is intentional, inclusive, systematic, and includes all aspects of the organization and community.

Further Discussions

- Define sustainability and resilience and the importance of both.
- Discuss the differences between sustainability planning and resilience planning.
- Explain the history and evolution of sustainability and the next phase of sustainability and resilience planning.
- Analyze the critical characteristics of sustainability efforts and sustainable and resilient organizations.
- Assess the climate change risks to local governments and their effects on operations and service delivery.

References

Adams C, Frost G, Webber W (2012) Triple Bottom Line: Review of the literature. In: Henriques A, Richardson J (eds) The triple bottom line: Does it all add up? Assessing the sustainability of business and CSR. Earthscan, London\Sterling, pp ix–xxii

Alibašić H (2013) The evolution of sustainability planning in grand rapids. Triplepundit People, Planet, Profit. Retrieved from http://www.triplepundit.com/2013/11/evolution-sustainability-planning-grand-rapids/?doing_wp_cron=1384104746.5217690467834472656250

Alibašić H (2017) Measuring the sustainability impact in local governments using the quadruple bottom line. Int J Sustain Policy Pract 13(3):37–45

Alibašić, H. (2018a) Ethics and Sustainability in Local Government. In Farazmand, A. (Ed.) Global Encyclopedia of Public Administration, Public Policy, and Governance. Springer International Publishing. https://doi.org/10.1007/978-3-319-31816-5_3427-1

Alibašić H (2018b) Role of corporations in addressing climate change. In: Farazmand A (ed) Global encyclopedia of public administration, public policy, and governance. Springer International Publishing, Zürich. https://doi.org/10.1007/978-3-319-31816-5_3427-1

Ammons DN, Smith KW, Stenberg CW (2012) The future of local government: will current stresses bring major, permanent changes? State and Local Government Review 44(1):64S–75S. https://doi.org/10.1177/0160323X12454143

Beeke C (2010) Grand rapids named most sustainable city in nation. MLive. Retrieved from http://www.mlive.com/business/west-michigan/index.ssf/2010/05/post_30.html

Bendell J, Kearnis K (2005) The political bottom line: the emerging dimension to corporate responsibility for sustainable development. Bus Strateg Environ 14(6):372–383. https://doi.org/10.1002/bse.439

Boomgaard J (2011) Making progress: Grand Rapids updates sustainability plan after meeting targets early. Mi-Biz. Retrieved from http://www.mibiz.com/news/economicdevelopment/18394-making-progress-grand- rapids-updates-sustainability-plan-after meeting-targets-early.html

Brownsword RA, Fleming PD, Powell JC, Pearsall N (2004) Sustainable cities – modeling urban energy supply and demand. Appl Energy 82:167–180. https://doi.org/10.1016/j.apenergy.2004.10.005

City of Grand Rapids (2009) Energy Efficiency and Conservation Strategy (EECS).

City of Grand Rapids (2011) Fiscal plan FY 2012–2016. Fiscal services. Retrieved from https://www.grandrapidsmi.gov/files/assets/public/departments/fiscal-budget/budget-office/fy11-final-fiscal-plan.pdf

City of Grand Rapids (2016) Sustainability plan FY2017–FY2021. Office of energy and sustainability. [Alibašić H (ed)] Retrieved from https://www.grandrapidsmi.gov/Government/Sustainability

Community Sustainability Partnership (CSP) (2017) Community sustainability partnership endorsing partners. Retrieved from: http://grpartners.org/about/endorsing-partners/

Coyle SJ (2011) The built environments and its supporting systems. In: Coyle SJ (ed) Sustainable and resilient communities: a comprehensive action plan for towns, cities, and regions. John Wiley & Son, Inc., Hoboken, pp 1–13

Elkington J (1997) Cannibals with forks: the triple bottom line of 21st century business. Capstone, Oxford

Elkington J (2012) Enter the triple bottom line. In: Henriques A, Richardson J (eds) The triple bottom line: does it all add up? Assessing the sustainability of business and CSR. Earthscan, London/Sterling, pp 1–17

Elster J, Hylland A (eds) (1989) Foundations of social choice theory. Cambridge University Press, New York

Fitzgerald J (2010) Emerald cities: urban sustainability and economic development. Oxford University Press, New York

Gaertner W (2009) A primer in social choice theory (revised edition). Oxford University Press, New York

Galea C (ed) (2004) Teaching business sustainability: from theory to practice. Greenleaf Publishing, Sheffield

Geary C (2011) Sustainable connections: linking sustainability and economic development strategies. National League of Cities (NLC). City practice briefs. Retrieved from http://www.nlc.org

Girardet H (2006) Creating sustainable cities. Schumacher briefings. Green Books. (Original work published 1999), Devon

Goodwin D (2012) Grand Rapids, Michigan: linking sustainability to performance. The International City/County Management Association (ICMA). Center for Performance Measurement™. Retrieved from http://www.icma.org/performance

Hardjono T, Van Marewijk MV, de Klein PD (2004) Introduction to the European corporate sustainability framework (ECSF). J Bus Ethics 55(2):99–113. https://doi.org/10.1007/s10551-004-1894-x

Heal GM (1998) Valuing the future: economic theory and sustainability. Economics for a sustainable earth series. Columbia University Press, New York

Hess DJ, Banks DA, Darrow B, Datko J, Ewalt JD, Gresh R, Hoffmann M, Williams LDA (2010) Building clean-energy industries and green jobs: policy innovations at the state and local government level. Science and Technology Studies Department, Rensselaer Polytechnic Institute, Troy. Retrieved from http://www.davidjhess.org/BuildCleanEnergyReportText.pdf

Hitchcock DE, Willard ML (2008) The step-by-step guide to sustainability planning: how to create and implement sustainability plans in any business or organization. Earthscan, London

IPCC (2014) Climate change 2014: synthesis report. Contribution of Working Groups I, II and III to the 5th Assessment Report of the Intergovernmental panel on climate change [Core Writing Team, RK Pachauri and LA Meyer (eds)]. IPCC, Geneva, Switzerland, p 151

Ismail S, Malone MS, Geest YV (2014) Exponential Organizations. Why new organizations are ten times better, faster, and cheaper than yours (and what to do about it). Diversion Books, New York

Knapp D (2011) Grand Rapids emphasizes climate adaptation in its sustainability plan. ICLEI USA. Retrieved from http://www.icleiusa.org/blog/archive/2011/07/06/grand-rapids-emphasizes-climateadaptation-in-its-sustainability-plan

Martin LL, Levey R, Cawley J (2012) The "new normal" for local government. State Local Gov Rev 44(1):17S–28S. https://doi.org/10.1177/0160323X12440103

McCarty K (2012) Beaverton, Grand Rapids win top Climate Protection Award. U.S. Mayor, 79(9)

Patton MQ (2002) Qualitative research and evaluation methods, 3rd edn. Sage Publications, Inc., Thousand Oaks

Portney KE (2012) Our green cities. Taking sustainable cities seriously. Retrieved from http://www.ourgreencities.com/

Saha D (2009) Factors influencing local government sustainability efforts. State and Local Government Review 41:39–48. https://doi.org/10.1177/0160323X0904100105

Savitz AW, Weber K (2006) The triple bottom line: how today's best-run companies are achieving economic, social, and environmental success-and how you can too. John Wiley and Sons, Inc., San Francisco

Slavin MI (2011) The rise of the urban sustainability movement in America. Creating the green metropolis. In: Slavin MI (ed) The rise of the urban sustainability movement in America. Island Press, Washington, DC

Solow RM (1992) An almost practical step toward sustainability: an invited lecture on the occasion of the fortieth anniversary of Resources for the Future. Resources for the Future. Island Press, Washington, DC

Steiner A (2017) The road to 100: Grand Rapids' journey to be Michigan's first all-renewable-powered city. Rapid Growth. Retrieved on Aug 17 2017, from http://www.rapidgrowthmedia.com/features/GRrenewable100.aspx

Stubbs W, Cocklin C (2008) Conceptualizing a sustainability business model. Organization Environ 21(2):103–127. https://doi.org/10.1177/1086026608318042

Stukkie H (2012) The greening of Grand Rapids. Rapid Growth. Retrieved from http://www.rapidgrowthmedia.com/features/08162012grgreen.aspx

The National Aeronautics and Space Administration (2016) Global climate change. [Online]. Retrieved from http://climate.nasa.gov/effects/

The U.S. Climate Change Science Program and the Subcommittee on Global Change Research. (USCCSP) (2009) Synthesis and assessment product 4.1 Report by the U.S. climate change science Program and the Subcommittee on global change Research. Retrieved from: http://www.globalchange.gov/sites/globalchange/files/sap4-1-final-report-all.pdf

The U.S. Global Change Research Program (USGCRP) (2014) The 2014 National Climate Assessment. Retrieved from: http://nca2014.globalchange.gov/downloads

The United Nations Framework Convention on Climate Change (UNFCCC) (2014) Kyoto protocol. Retrieved from http://unfccc.int/kyoto_protocol/items/2830.php

United Nations General Assembly (1992) Report of the United Nations Conference on Environment and Development. Rio de Janeiro, 3-4 June 1992. Annex I Rio Declaration for environment and Development. Retrieved from: http://www.un.org/documents/ga/conf151/aconf15126-1annex1.htm

United Nations University (UNU) (2017) Institute for the Advanced Study of Sustainability. Retrieved from: https://ias.unu.edu/en/

United Nations, World Commission on Environment and Development (1987) Our common future. Oxford University Press, Oxford. Retrieved from http://www.un-documents.net/wced-ocf.htm

Vande Bunte M (2012) Inside the numbers in Grand Rapids: Food trucks, budget worries and a national energy honor. MLive. Retrieved from http://www.mlive.com/news/grand-rapids/index.ssf/2012/06/inside_the_numbers_in_grand_ra.html

von Bertalanffy L (1950) An outline of general system theory. British J Philos Sci 1(2):134–165. https://doi.org/10.1093/bjps/I.2.134

Waddock S (2009) Leading corporate citizens: vision, values, and value added, 3rd edn. McGraw-Hill Irwin, New York

Werbach A (2009) Strategy for sustainability: a business manifesto. Harvard Business Press, Boston

West Michigan Environmental Action Council (WMEAC) (2013) Grand Rapids Climate Resiliency Report. Retrieved on Sept11, 2017, from: https://wmeac.org/wp-content/uploads/2014/10/grand-rapids-climate-resiliency-report-master-web.pdf

Wood DJ (2009) City of Grand Rapids named one of EPA's top 20 'green' energy purchasers nationwide. The Rapid Growth. Retrieved from http://www.rapidgrowthmedia.com/innovationnews/CityEPA0219.aspx

2 Mapping Out the Sustainability and Resilience Process for Organizations and Communities

"To the multiple valorizations of wild environments can be added mystery. Without mystery life shrinks. The completely known is a numbing void to all active minds. Even a laboratory rat seeks the adventure of the maze." Page 146, The Future of Life, Edward O. Wilson

Key Questions

The second chapter of the book is aimed at answering the following underlying assumptions and questions:

- Does an organization have a strategic plan, and what strategies are currently in place that may be viewed as sustainability- and resilience-related?
- What are the long-term goals, vision, values, and core mission of an organization pursuing sustainability and resilience?
- What sustainability and resilience programs does the organization have in place and how are they defined?
- What are the substantive sustainability and resilience efforts in the organization?
- Does the organization have a plan to reduce greenhouse gas emissions or to minimize the impact of climate change?
- Does the organization have a plan for recycling, energy efficiency, renewable energy, and waste minimization?
- What is the current budgetary and financial situation in an organization?
- What are the available internal resources, the capacity of the organization?
- How does an organization address its long-term planning; strategies for economic growth, social justice, and equity; environmental protection; good governance; and resilience?

Introduction

Building on the first chapter, the second chapter outlines and defines processes in determining the appropriate steps and resources in sustainability and resilience planning for organizations. The second chapter is intended to provide answers to the queries related to an existing strategic plan, long-term goals, vision, values, and core mission of an organization pursuing sustainability and resilience planning. It enables the readers to understand the external and internal dynamics of organizational commitment to sustainability and resilience, threats, and opportunities and to evaluate them in the context of budgetary and financial circumstances surrounding the organization. This chapter maps out the sustainability and resilience planning process for organizations and communities, determining the appropriate steps to be taken at each level of sustainability and resilience planning. The sustainability planning process includes the environmental scan, mapping out the current conditions and available internal resources, the capacity of the organization, and an analysis of opportunities that may exist in the community, region, state, and the nation. The purpose of the second chapter is to provide organizations with tools to recognize resources available within the organization, coupled with challenges and opportunities. It also

H. Alibašić, *Sustainability and Resilience Planning for Local Governments*, Sustainable Development Goals Series, https://doi.org/10.1007/978-3-319-72568-0_2

enables the leadership to understand and comprehend sustainability efforts already underway within organizations.

Leadership within an organization starts with vision and mission statements, review of current conditions and existing sustainability efforts, long-term goals, and the budgetary/financial condition of the organization. Many organizations, including local governments, have sustainability efforts in existence through recycling, composting, energy efficiency, waste management, bike lanes and biking programs, watershed protection, renewable energy, and other sustainability-related activities. However, often, organizations are not aware of the implications of sustainability and resilience activities or do not measure corresponding outcomes. A proposed approach to sustainability planning would be to use the Quadruple Bottom Line approach. Additionally, in light of climate change threats, many local governments develop climate action plans, enhanced emergency preparedness strategies to account for climate change threats, resilience plans, and climate mitigation and adaptation strategies. A thorough review of the existing plans is warranted, including the master plan, traffic safety plans, emergency preparedness plans, hazard mitigation plans, and other relevant programs, policies, and procedures, including the cities' budget and fiscal policies.

Sustainability Plan Staging and Quadruple Bottom Line (QBL)

Aligning the sustainability planning with the fiscal year, budget process and defining the benchmarks and baseline year is critical. Moreover, aligning benchmark data and targets with the existing plans and policies is essential to the longevity of sustainability and resilience in organizations. The plan needs to identify evolving sustainability priorities; guiding principles; key objectives and strategies in an era of organizational, community-wide, and climate-related initiatives; and transformation efforts. The goals and targets need to be outlined in the sustainability plan to equip local government officials to respond to future obstacles and opportunities in a viable and coordinated manner. The sustainability plan acts as an adjunct to various city's plans and ideally expands the Triple Bottom Line framework into a Quadruple Bottom Line (QBL) structure. Quadruple Bottom Line (QBL) serves as the overarching themes of this plan: economic, social, environmental, and governance. The specific goals of a plan may be housed under one of the four themes, and specific targets are categorized under ten separate goals: economic opportunity, great neighborhoods, social equity, safe community, resilient systems, balanced transportation, sustainable assets, fiscally resilient, transparent and accessible, and good government.

For cities that already have an existing plan in place, the new sustainability plan expands on the previous ones. Most of the plans rely on the Triple Bottom Line (TBL) framework. However, a Quadruple Bottom Line (QBL) structure allows for governance and related efforts to be appropriately addressed in the sustainability plan. Also, drawing from the city's existing plans and aligning them with its goals, outcomes, and themes allow for consistency, continuity, and integrity of operations. The Quadruple Bottom Line provides the overarching pillars of this plan: economic, social, environmental, and governance. The potential themes of a citywide sustainability plan may be housed under one of the four QBL pillars. Each specific target is categorized under separate goals and outcomes, directly connected to the city's budgetary and fiscal plans:

1. Economic growth and opportunities
2. Resilient environment
3. Resilient and safe neighborhoods
4. Resilient assets: infrastructure, buildings, utilities, and balanced transportation
5. Good governance: accountability, transparency, accessibility, community input, and fiscal resilience

Accountability, transparency, and accessibility are vital to ensuring communities and city governments are more sustainable and resilient. Organizations, citizens, institutions, and businesses and decisions, actions, and activities they undertake have an impact on the overall resilience and sustainability of a community. It is the responsibility of appointed officials and employees to contribute to the implementation of sustainability targets. Furthermore, staff assigned to be the target champions are responsible for leading and achieving specific sustainability targets outlined in this plan. This approach enables individual accountability and ownership for each goal and its effectiveness.

Organizations use a sustainability and resilience plan to further good governance and improve the operational efficiency. The plans can be utilized to address affordable housing opportunities, diversity, and inclusiveness, decrease energy consumption, intensify renewable energy production, minimize waste, support community outreach efforts, standardize the system processes, and partner with organizations, companies, and citizens to further spur growth while preserving natural resources. Local government employees track the progress of each target and create a public record outlining both the plans and actions that address the Quadruple Bottom Line (QBL). The sustainability and resilience progress reports are released to the public annually or biannually. Cities alone, however, cannot implement all of the strategies outlined in the plans and strive for a partnership with other organizations within community, region, or at state and national stage.

Strategic vs. Sustainability Planning in Cities

Certain local governments continue to utilize both the strategic plan and sustainability plan. An ideal situation would be to use the sustainability plan, with a set of specific, well-defined Quadruple Bottom Line targets. Figure 2.1 describes the process and the benefits of choosing the sustainability plan within an organization. Setting a defined, target-driven plan in place with a set of measurable outcomes enables local government administrators to deal better with ongoing demands for services, achieve long-term goals, and define a long-term vision for the future. Beyond simple strategic components, a well-defined sustainability and resilience planning effort enable consistency in service delivery, continuity of operations, and execution of long-term goals and objectives.

For instance, Michigan Meridian Township, which has a history of environmental stewardship, developed a climate sustainability plan, combining climate resilience and sustainability planning into one. By taking such approach, local government can constructively coalesce its planning efforts into a single visioning and guiding document, ensuring consistency in planning and reporting. As noted in the climate sustainability plan, the intended implementation goal of the plan is for Meridian Township to "contribute to the worldwide efforts to curb greenhouse gas emissions" and make the "community a more sustainable, resilient, affordable, and vibrant place to live" and to make "government operations more energy and resource efficient and better prepared to deal with the impacts of climate change" (Meridian Township 2017, pp. 2 & 16). In addition to listing the efforts of the township in addressing sustainability challenges to date, the authors of the report noted the township's commitment to Paris Climate Accord.

In response to frequent threats exacerbated by climate change and extreme weather, in addition to sustainability plans, cities formulate a climate action plan too. Boswell et al. (2012) claimed the climate action plans to be "strategic plans that establish policies and programs for reducing (or mitigating) a community's greenhouse gas (GHG) emissions and adapting to the impacts of climate change," seeing them as "visionary, setting broad outlines for future policy development and coordination, or they may be focused on implementation with detailed policy and program information" (p. 67). While they are undoubtedly essential documents, the scope and approach to resilience planning would include both the sustainability and climate action plan, with a broader view on the community's sustainability, climate adaptation, mitigation, and preparedness strategies. Climate action strategies and plans may be integrated into an existing sustainability or a resilience plan.

Sustainability as policy and programmatic framework is employed by organizations in both the private and the public sector. While there are

Sustainability Plan

- Ongoing, progressive, specific targets with baseline year and benchmarks.
- Continuous events.
- Narrowing focus to include climate change/resilience/ traditional sustainability
- Most cities focusing on mobility, water, food systems, energy, infrastructure.
- Broader set of specific, dynamic goals.

Strategic Plan

- One-time, event oriented change.
- Less long term focused.
- Discrete events/ more direct.
- Similar in structure. However, economic, social, environmental, and governance goals, targets or indicators not called out or specified.

Concerns

- Duplicating efforts unnecessarily.
- Plans meet the same organizational objective?
- Too similar in structure, but written very differently.

Solution

1) Focus: Define the Quadruple Bottom Line opportunities, with measurements in the area of environmental, social, economic, and governance targets.
2) Use sustainability planning to address strategic concerns, long term vision, mission, and goals with specific targets.

Fig. 2.1 Sustainability vs. strategic plan

conflicting opinions on the definition of sustainability, its impact on organizations, and the practicality and applicability of sustainability, the clarity of sustainability and resilience planning is embraced by many local governments. More importantly, existing sustainability and resilience planning enables the city administrators to apply the practical elements of Quadruple Bottom Line to its operations and service delivery. Sustainability and resilience planning is considered to be an innovative, original, and novel approach employed by organizations. Focusing on the organizational effectiveness and efficiency through the successful application of sustainability and resilience planning provides a better understanding of the practical implications of sustainability.

Cumo et al. (2012) implied in their research that "urban areas are in fact the places where the on-going transformation of environment, society, economy, and their complex impacts become concrete, need to be managed and must be taken into consideration for the present and the future generations" (p.29). When evaluating urban areas and their negative environmental impact, the issue of the exact definition of what constitutes the urban area comes into play. Kennedy and Sgouridis (2011) argued the difficulty of delineating the exact urban boundaries when determining the greenhouse gas emissions impact and how "this interconnectedness complicates the task of determining which emissions should be included in a city's" overall account of greenhouse gas emissions (p. 5261).

There are multiple steps involved in mapping out sustainability and resilience planning in organizations and then its auxiliary impact on a community. The first step to sustainability planning is defining a mission and vision statement, aligning them with the long-term organizational goals, determining the type of activities to be measured and what level, defining stakeholders, and prioritizing areas of responsibilities. The city's sustainability and resilience plan, while it relates to the city as an organization with its operations and resources, also focuses on the broader implication of sustainability in the region and impact on the community. A sustainability plan serves as a guidance for organizational strategic initiatives. The sustainability plan serves as a long-term strategy. However, a more comprehensive plan relying on climate science and the study of extreme weather,creates a basis for the resilience plan.

Mapping Out Current Conditions and Resources

The external pressures on the built environment are best described using the Quadruple Bottom Line approach. The demographic changes and trends, the income level of the city population, and socioeconomic movements fall under the social and economic categories. Environmental factors include the quality of water, air, available natural resources, industrial pollution, and other determinants impacting the quality of the environment. Leadership plays an important role, including encouraging community engagement, accountability, transparency, fiscal responsibility, answerability, ethics, and integrity of the public service. Droege (2006) argued the importance of urban areas and cities in addressing the increase in fossil fuel-related emissions. Seeking sustainability-driven policies and practices in cities as they "are the most advanced but also the [riskiest] and fragile constructs ever devised by humankind" and therefore must be more proactive in addressing the risks of greenhouse gas emissions (Droege 2006, p. 142).

Moreover, Lindfield posited that (2010) "globalization and restructuring of national economies have resulted in the outsourcing of production and services" (p.108). In essence, this shift in economic growth created new challenges for cities, as urban centers became centers of economic activities and related adverse environmental impact on cities. On the other end of the spectrum, localities in the United States and Canada tackled the loss of manufacturing jobs amid a shift to new industries and outsourcing of production, without adequate infrastructure investments, and lack of support for education, to name a few. Sustainability and resilience plans are used as planning and strategy tools and policies adopted on local, national, and international levels for cities to successfully combat increasing threats and address risks of unsustainable planning policies and practices.

Cities are significant contributors to the overall greenhouse gas emissions (GHG). Measures to counteract the effects of GHG and global warming include energy conservation; waste minimization supported by reusing, repurposing, and recycling of materials; public transit; bicycling; pedestrian-friendly neighborhoods; quality of life; and a cleaner and greener community. Cities are continually seeking strategies to promote the integrity of the natural environment, including energy use decrease, climate protection, improved environmental quality and natural systems; sustainable land use, urban form, and transportation. By positioning sustainability and resilience planning through the Quadruple Bottom Line, local governments adequately and appropriately address ongoing threats. Furthermore, cities embrace opportunities, expressed in emergent practices, such as renewable energy production, electric vehicles, charging stations, tree planting, increasing the availability of affordable housing options, and addressing social equity and other issues (Fig. 2.2).

Sustainability and resilience planning enables organizations to use a multifaceted, cross-sectoral approach for the betterment and operational efficiency of organizations. The primary drivers for successful sustainability programs are the needs and desires of communities and organizations to change and to adapt to the changes in the environmental, societal, and economic conditions surrounding them. Thorough sustainability and resilience planning enables organizations to create a transformational culture, allowing staff to embrace the mission and values of everyone for an organization. Moreover, sustainability and resilience enable administrators to integrate all elements of the system and holistically consider them.

Sustainability and resilience planning requires a detailed overview and analysis of the current conditions in the organizations, including a thorough review of risks and opportunities in the environment. Developing an accurate sustainability plan goals and objectives should at a minimum

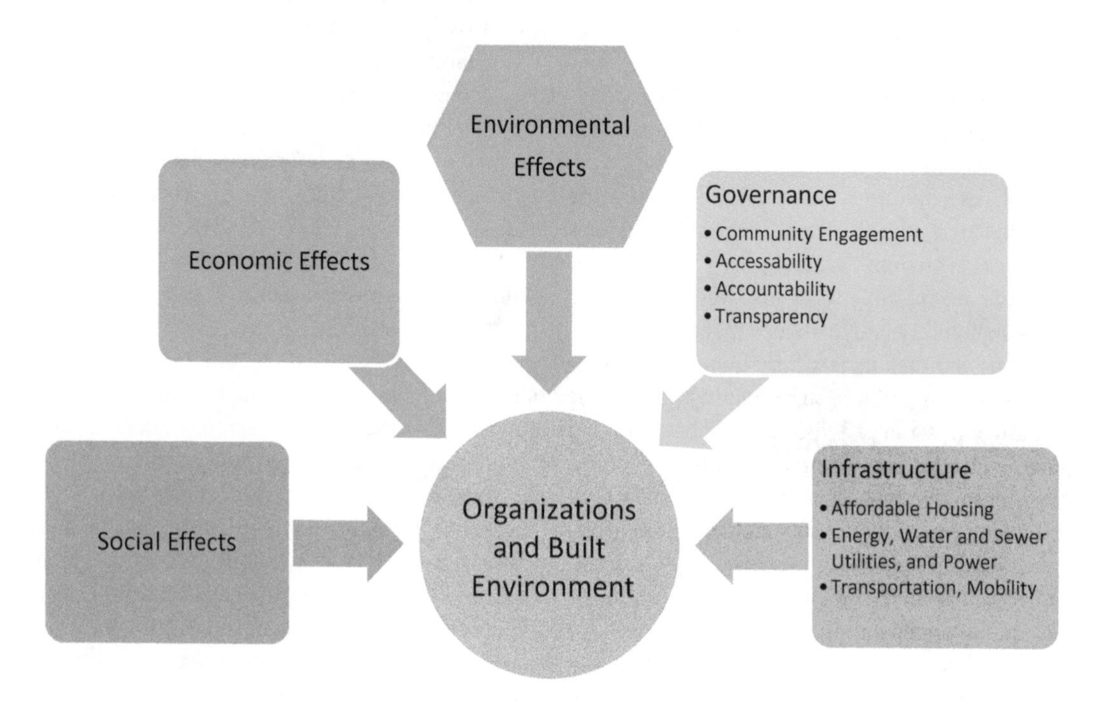

Fig. 2.2 The QBL and external pressures on organizations within built environment

include a vision statement and long-term and short-term goals and identify potential targets that the city could adopt. For example, various sustainability planning strategies involve conducting assessments, creating indicators, writing a plan, implementing initiatives, and reporting outcomes. However, more in-depth strategies and steps involved in the preliminary design of the sustainability plan should at a minimum include the following steps in the process:

1. The first step is a detailed analysis of the vision, mission, and goals, with broad objectives from long- and short-term planning perspective. Other documents such as master plan, traffic safety plans, parks and recreation plan, emergency preparedness plans, hazard mitigation plans, strategic plans, climate plans, and other reports need to be reviewed for alignment. A scan and analysis of the city's relevant documents, the past and current sustainability-related practices and policies, are a must for effective sustainability planning to frame preliminary recommendations for a sustainability plan. Select interviews with the elected officials and appointed officials need to be included in this process.
2. The second step includes a review of the budget, fiscal conditions, economic growth trends, and social demographics, including population trends, development patterns, and commercial and residential housing demands. The step needs to include a thorough understanding of the current budget. A city may have a 100% renewable energy target. However, cities need to take into consideration any associated costs with a switch to cleaner energy sources, including vehicles, equipment, or powering buildings and operations.
3. A thorough review of the existing energy use by the entire organization, including energy consumption and, in the case where cities own a power utility, energy production. Developing a detailed review of greenhouse gas emissions, and defining the outline for the emission target reductions as a result of the energy use and energy production. The purpose of the GHG inventories is to provide a baseline against a particular benchmark or base year so that activities may more accurately measure progress toward the reduction of its emissions. GHG emissions are represented in metric tons of carbon dioxide equivalent (CO2e) produced by energy consumption and other activities of the organization and the community.

   The GHG inventory is customarily divided into three subcategories, such as direct city's operations, employees' caused emissions, and community-wide emissions. Each category is broken down into direct and indirect emission, with baseline year and the year from which data was collected. With a myriad of sources of data for energy use, demographics, and types of emissions, it can be time-consuming and confusing for cities to conduct the carbon footprint inventory. A more in-depth explanation for GHG emission or carbon footprint inventory is provided in the Tools and Resources Book Chap. 7.

4. Existing human resources; staff and supporting staff; community-wide resources, such as nonprofit, health, and human services; and community facilities.
5. A review of existing infrastructure, infrastructure needs and plans, capital projects, including transportation, mobility, roads, sewer and water utilities and facilities, power, and green infrastructure.
6. A review of environmental programs, water and air quality, ozone day programs, waste management, recycling efforts, river and waterways clean up, and other practices, programs that encourage environmental practices.
7. Finally, in the preparatory stage of the sustainability plan development, a scan and an analysis of the governance are in order. Additionally, a review of the local governments' public engagement, website information, announcement, communication strategies, neighborhood participation programs, partnership opportunities, and ongoing community networks.

Sustainability and resilience planning launches a method to identify the current state of the organization, where it is going, and how it will get there. This scanning stage encourages the consistent decision-making, communication, and performance assessments and can create a sense of cohesiveness with the organization. In embracing these elements, organizations create a building block to ultimately enhance the outputs of the organization.

Sustainability "Plan-Do-Check-Act" Cycle

Sustainability planning management is a never-ending process that combines strategic planning and leadership with other management processes. As noted in Alibašić (2017), sustainability "may be defined as a set of effective and efficient actions taken by an organization, through good governance, to ensure the economic stability, growth, and financial success, with the most positive societal outcome and the least negative environmental impact" (p. 37). As such, in sustainability and resilience planning, organizations must take into consideration all four components and integrate them effectively into their long-term goals and objectives. The measurement of sustainability and resilience through regular reporting mechanisms allows organizations to track progress and enables them to use measures as a tool for refinements and loop. Similar to Lean, Plan-Do-Check-Act cycle, for planning, implementing, feedback, and reapplying and refining processes, sustainability and resilience planning has a never-ending loop function, intended for continuous improvement.

It is imperative for an organizational leadership team to use sustainability planning in response to the rapid changes occurring in the surrounding environment. Sustainability planning must be viewed as dynamic, ever-changing, all-hands-on-deck approach to every level of organizational governance, employed and deployed with maximum resilience in mind. In this light, the process of sustainability planning may be viewed as a force allowing organizations to substantially address the critical planning and avoid a crisis as a situation requires and on any scale. Sustainability planning is a forward-thinking system of techniques, which involves evaluating and analyzing developing practices, opportunities, or perils to the organization and developing a comprehensive response, taking advantage of the existent internal and external resources. Organizations can respond to threats in a purposeful mode, with resilience and elasticity as ultimate goals.

Sustainability and resilience planning allows the executives and staff work together and to better enhance the fulfillment of the mission, goals, vision, and meeting of mandates, with continuous improvements, and the sustained performance. This package of structural definition allows an organization to function effectively. Similar to strategic planning, sustainability planning can start with the three what questions: what is the strategic position of the organization now, what does the organization strive to accomplish and achieve, and what resources does the organization have at its disposal to meet the goals and objectives. Answering specific sustainability "what" questions will conclusively lead to a change and path forward in mission, vision, and goals. While historical crises, such as financial recessions and resource shortages, recur with vigor, new challenges and opportunities, including but not limited to artificial intelligence, social media, and cyber warfare, have emerged to present nascent leaders with a more significant deal to contend with challenges of the contemporary, fast-paced, and technology-reliant world.

As local governments face continual scrutiny, sustainability initiatives provide administrators with new instrumentation to guide their organizations to a more predictable future with clear goals and understanding of purpose. The Quadruple Bottom Line assists organizations in establishing a path forward and a sense of direction for the organization. An essential aspect of sustainability and resilience planning as it relates to public sector organizations is the nurturing of the transformational and innovation culture within an organization, which links planning and implementation and continually challenges the notion of status quo. The Plan-Do-Check-Act (P-C-D-A) cycle applies specifically to public service organizations, to foster an atmosphere and organizational culture where all members are focused on carrying out the mission and values of the organization with an understanding of the direction the organization is headed. The example of Plan-Do-Check-Act cycle from the city of Grand Rapids illustrates the significance of direct loop and feedback between the initial sustainability plan and the implementation, with the return on investment from sustainability initiatives connecting to the city's budget annually.

The P-D-C-A cycle enables the connection of and the assessment of existing plans, allowing organizations to measure progress using a sustainability or resilience plan. In this case, sustainability plan is not only used as a conduit but also as a report on the overall outcomes and investments made by the organization. The P-D-C-A cycle and Lean principles are efficiently fused with core operations, ultimately resulting in savings and process improvements (Fig. 2.3).

Resilient City Spotlight: Resilience of Local Communities, New Orleans

In 2015, the city of New Orleans released its first resilience strategy, which outlines steps for making and building a more resilient New Orleans. The process of resilience planning took years. Among many of the steps undertaken by the city of New Orleans was a development of the city of New Orleans Carbon Footprint Report. The report built on the previous carbon footprint reports, including 2001 report, with the baseline year of 1998. Very importantly, the city recognized the need to reset the carbon footprint data after the devastating effects of Hurricanes Katrina and Rita in 2005. As noted in the 2009 report, climate change and projected sea level rise continued to pose significant threats to the city unless current rates of greenhouse gas emissions are drastically curbed and reduced efficiently.

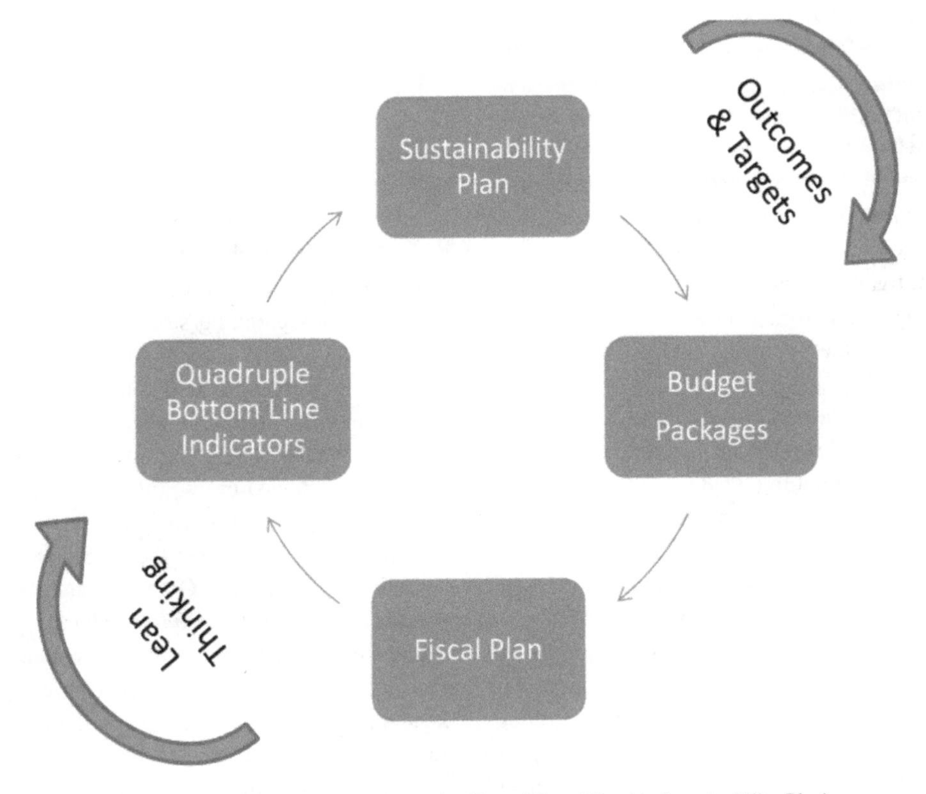

Fig. 2.3 Plan-Do Check-Act (PDCA). (Adapted from the City of Grand Rapids Sustainability Plan)

New Orleans' leaders recognized the disproportionate impact, vulnerabilities of the community, and the effects that global climate change has on the city. The report pointed out the impacts and threats of increased frequency and intensity of hurricanes, including the ultimate negative consequences to the city and the region, such as "higher prices and shortages of basic goods, such as food and energy; increased public expenditures on relief and rebuilding due to extreme weather events; a higher susceptibility to flooding; and a higher rate of infectious diseases and heat-related illnesses and deaths" (City of New Orleans 2009, p.4). Cities with unique threats must learn to adapt and mitigate the impact of their operations on the environment and the society. Consequences of not doing anything to reduce the carbon footprint are dire. In New Orleans alone, if "no actions were taken to reduce GHG emissions from the date the data were collected, the City of New Orleans would produce 6,451,399 tons CO2e in the year 2030, a 36.5% increase from 2007" (City of New Orleans 2009). Some of the adaptation and mitigation activities included in the New Orleans master plan are the rehabilitation and enlargement of the city's urban forest with a specific benchmark for a percentage of the citywide tree canopy by a certain year. Given the extent of the damage to the trees from hurricanes, this target addressed the adaptation and mitigation strategies. Besides, the city encourages and promotes tree planting, preservation on both the publically and privately owned properties.

Given the vulnerabilities to storms and loss of land area, New Orleans is particularly interested in establishing sustainable stormwater management practices and protecting environmentally sensitive areas, such as wetlands, from adverse impacts to enhance the city's water-storage capacity during storms and increase protection against storm surges. In more recent years, and with the development of city's resilient strategy, there is further recognition of the unending impact of climate change and persistent threats that exist in the coastal cities. In the Resilient New Orleans report, in addition to the resilient strategies for regional transportation, promoting sustainability as a growth strategy, reducing redundancy and improving reliability of the power supply, integrating resilience into decision-making process, and investing in pre-disaster planning and post-disaster recovery, the city focuses on developing strategies and planning for resilience at neighborhood and business district level (City of New Orleans 2017). In looking at all the relevant aspects of resilience, the city also emphasized the issue of equity as an essential component to resilience:

> "Even as we look to the future, we cannot ignore past injustices. Racial inequity is present in every facet of our society- employment and income, education and health, violence and justice, housing and social mobility. To advance as a city, we must confront this reality collectively and seek meaningful ways to address its effects in our institutions, our communities, and our families. With a strategy that prioritizes racial equity, we will be stronger as a society and more capable of responding to adversity" (City of New Orleans 2017, p. 11).

Elkington (1997) defined sustainability regarding the "triple bottom line focusing on economic prosperity, environmental quality and the element which business has tended to overlook – social justice" (p.2). Social justice is often overlooked in local government sustainability planning. Defining it through the prism of a strategic resilience planning strengthens the organizational vision, mission, and goals.

Key to the success of an organization's attempt to sustainability and resilience planning is its ability to express its vision and goals for such preparation transparently. Integrating steps to increase resilience into city's sustainability and other plans is imperative to the success of implementation of such plans. Organizations also need to provide a consistent and transparent reporting mechanism to the public and share results while also creating opportunities for participation and involvement. The years ahead will be able to show how well the city of New Orleans performed in building the resilience of the community.

Vision for Local Sustainability and Resilience Planning

As cities complete their preliminary review and scan of the sustainability and resilience inventory, efforts in place, opportunities, and threats, they need to proceed with defining a vision for sustainability and resilience. According to Coyle (2011), critical elements of sustainability include the built environment, energy, water supply, wastewater, stormwater, natural environment, transportation, food production, farming and agriculture, solid waste, economics, public engagement, and education (p.29). Moreover, Coyle (2011) astutely pointed out the value and relevance of the public and stakeholders engagement in the process. Resilience planning is more comprehensive and expansive, and it ensures that organizations and communities continue to operate, function, and thrive after experiencing a disaster. Cities need to define a vision for sustainability and resilience planning while accounting for all the elements of the Quadruple Bottom Line. Some of the elements considered in the sustainability and resilience visioning process within built environment include the following:

1. Protection of natural resources and the environment
   - Adopt aggressive water conservation and waterways protection measures.
   - Reduce direct discharges to rivers, tributaries, and oceans.
   - Promote pollution prevention and reduce toxic chemicals entering waterways, including pharmaceuticals.
2. Carbon reduction and clean energy economy
   - Reduce dependence on carbon-based sources of energy.
   - Promote renewable energy generation, from wind and solar and geothermal for heating and cooling.
   - Improve efficiency of transmission and hardening of the power grid.
   - Reduce energy consumption and waste.
   - Minimize waste and increase composting, waste reduction, repurposing, and recycling.
   - Promote efficient and low-carbon transportation alternatives and invest in public transit.
   - Support green job creation, research, and innovation, small- and medium-sized business.
   - Promote eco-friendly tourism and local economy.
3. Resilient built environment
   - Adopt low-impact, high-density urban planning with walkable communities.
   - Require energy efficient, green building planning and design.
   - Plan for green infrastructure to reduce stormwater including green roofs, permeable roads, and other measures to reduce water runoff.
   - Design and invest in efficient public transit.
   - Invest in green infrastructure.
   - Provide quality public safety, supporting first responders and public health agencies.
   - Create opportunities for affordable housing and investment in local business and neighborhood districts. (Fig. 2.4).
4. Good governance, accountability, answerability
   - Ensure fiscal responsibility, accountability, and responsiveness.
   - Engage public and educate about all the aspects of sustainability and resilience and seek continuous feedback.
   - Relate and report to the public all the progress and failures.
   - Engage staff and community organizations, neighborhood and business districts, nonprofits, universities, and health institutions.

Practical Approach: From Sustainability Planning to Climate Resilience

Cities are already experiencing climate change impacts in multiple ways that may include:

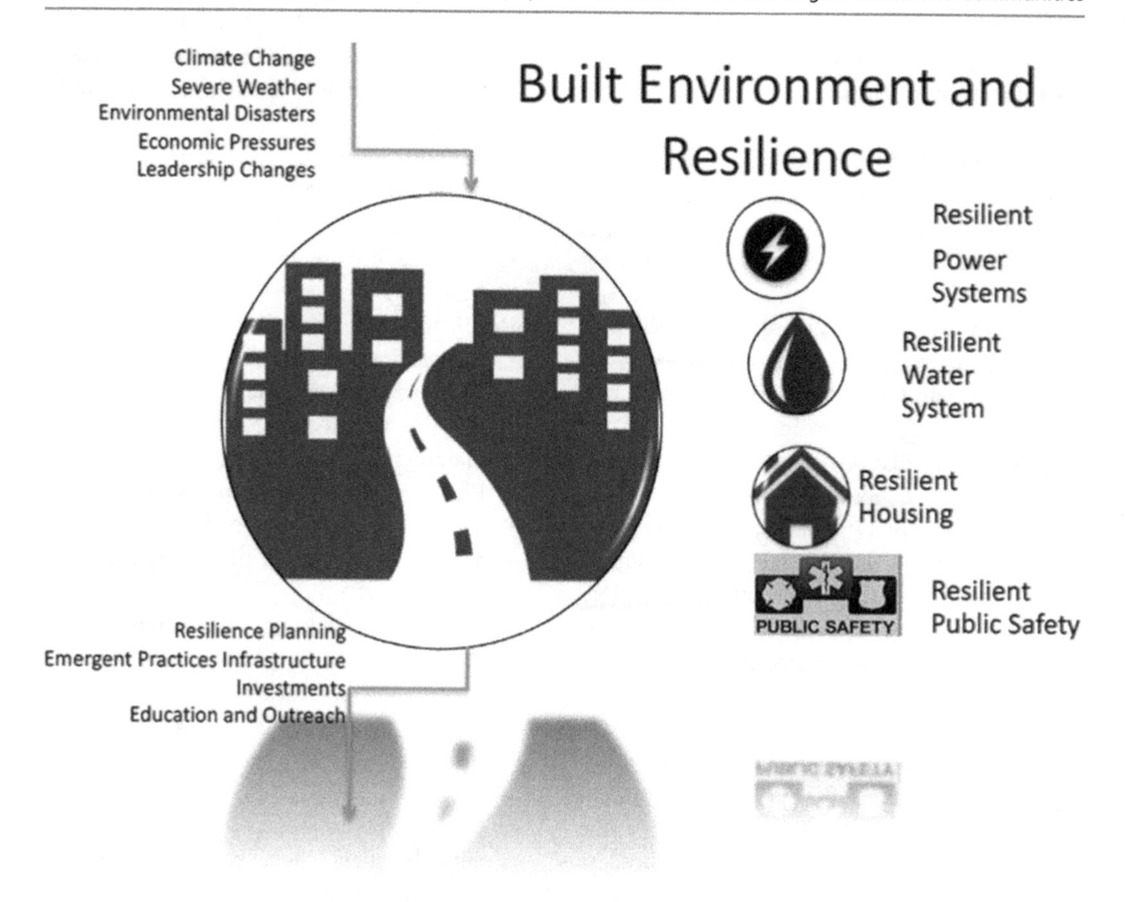

Fig. 2.4 Resilience and built environment

- Extreme heat waves putting the elderly, young, and socially disadvantaged at risk
- More frequent severe rain events, storms, and snowstorms that stress water, road, sewer, power, and other infrastructure
- Water shortages during frequent and intense droughts
- Increased smog, fire risk, and air pollution that exacerbate respiratory illnesses and other medical conditions
- Pests and disease risks

Using a system-wide approach to evaluate the climate change effects on the local community, West Michigan Environmental Action Council (WMEAC) in partnership with the city of Grand Rapids, developed the climate resiliency report with the localized climate change impact assessment. After over a thirteen months period of collaboration, interviews, and research, the report was first presented to the city commission and then made available to the public. The report was financed with a portion of the grant sponsored by Walmart for the city of Grand Rapids' 2012 US Conference of Mayors' Climate Protection Award, recognizing the city for sustainability planning and local climate protection efforts to reduce greenhouse gas emissions.

Notable aspects of the report included its focus on localization of climate change impact and a distinct list of suggestions for building resilience in the local community and strengthening disaster recovery and resilience of the local government. Notable goals of the resiliency report were to potentially initiate discussion and further enhance projects, policies, programs, and planning actions enabling Grand Rapids to mitigate the effects of climate change, to adapt to its impacts, and to utilize emerging sustainability opportunities (Alibašić 2013).

Local community partners in Grand Rapids also contributed to the report. Moreover, the report was built on a comprehensive set of 25 interviews of representatives from the public and public sectors from academic, first responders, regional planning, and utilities. Universities in Grand Rapids have partnered with the local government on sustainability and climate resilience planning for the community and the region. With the academic and research support from Grand Valley State University, the modeling software, Model for the Assessment of Greenhouse-gas Induced Climate Change with the Regional Scenario Generator (MAGICC/SCENGEN 5.2.3.) was utilized for analysis of Grand Rapids to the square area of 2.50 (175 miles) by 2.50 (175 miles).

System Approach and Localization of Climate Science

An essential element of combining resilience and sustainability planning is to take a systematic, holistic approach to organizational and community planning. As noted in the 2013 Grand Rapids climate resiliency report

> "Climate change impacts each sector in isolation, but it also impacts the interaction of each to others and the function of the system as a whole. Therefore, understanding the needs of the community, major relationships between sectors, and the ability of the sectors to provide those needs in a changing climate world is key to building resilience." (WMEAC 2013)

The report projected the climate change data of temperature and precipitation through the years 2022 and 2042, coinciding it with the city's 20-year master planning process. The annual and monthly baseline averages of temperatures and precipitation were compared to baseline data from 1961 to 1990. A sample of the analysis and findings from the report that describes climate change impact in Grand Rapids area:

- Average temperature and precipitation will increase by 1.1oC and 2.6%, respectively, by 2022, and further increase by 2.2oC and 8.5%, respectively, by 2042.
- Seasonally, the most substantial increases in temperature are projected to occur during the winter and the least in summer.
- The most significant percentage increase in precipitation is predicted to occur in the winter and spring months.
- Summer is the only season projected to become drier.
- The Great Lakes region can expect more variable and unstable weather. This volatile course could lead to more extreme weather events such as storms producing greater than 1 inch of rain in 24 h, increased recurrence of back-to-back days above 90° and 90% humidity, and more freeze-thaw cycles in winter and spring (WMEAC 2013).

Report Conclusions and Recommendations

The report includes conclusions and recommendations in the areas of process improvements as described in a sample of recommendations below.

- Under processes, organizations should use an economic, Triple Bottom Line cost-benefit approach in financing and implementation of notable projects. The city took a step further and adopted the Quadruple Bottom Line in its approach to sustainability.
- Under crime prevention, the authors of the report suggested the use of crime prevention tools through the environmental design of parks and public spaces and opening lines of communication with community and neighborhood organizations.
- The report concluded the city should seek to move from a centralized energy system toward a more distributed energy system, energy efficiency, and renewable energy systems.
- Continue to encourage the construction of best-in-class green building projects.
- Research and implement climate-resilient street maintenance and construction practices,

particularly for materials and physical infrastructure.

- Adopt a big urban tree canopy goal – at least 40% – and implement a forestry program addressing heat island, air quality, and other resiliency values delivered by a diverse, healthy urban tree canopy.
- Implement a forestry program addressing heat island, air quality, and other resiliency values delivered by a diverse, healthy urban tree canopy (WMEAC, 2013).

Future Work

The report was used to directly support and link to various aspects and targets of the city's sustainability plan. It provided an opportunity for regional and state-wide discussions on the impact of climate change but also for specific discussions on policies and tools to implement climate resiliency in communities and regions. As concluded in the report, Grand Rapids needed an individual or organizations to own and champion climate resiliency in the community. In a broader sense, the resiliency report itself may serve as a template for similar reports for other local and regional governments. As indicated by Boswell et al. (2012) "climate action plans are becoming the primary comprehensive policy mechanism for the reduction of greenhouse gas emissions and management of risks posed by climate change" (p. 49). The first step to a comprehensive resilience planning is a thorough and well-thought-out climate resiliency report.

Table 2.1 features recommendations from the Grand Rapids climate resiliency report, translated into actionable targets used in the sustainability plan. The process used to embed resilience targets into sustainability plan ensures the longevity of such strategic planning beyond current political leadership. Each target is then quantified and measurable.

Table 2.1 Climate resilience report recommendations

Climate resiliency report recommendations				
Water	Energy	Built systems/ infrastructure	Transportation	Emergency preparedness
Strengthen the water use efficiency Capture the first flush Use critical climate infrastructure	Increase energy efficiency Reduce GHG emissions	Improve access to food sources Increase the number of certified sustainable buildings	Change transportation culture to one built around multi-modal, and vital streets for all residents	Analyze the effectiveness of resources used during extreme events, continue providing efficient response
Corresponding sustainability plan targets				
Water	Energy	Built systems/ infrastructure	Built systems/ infrastructure	Emergency preparedness
Reduce customer water consumption Reduce stormwater discharge Increase square footage of green roofs, pervious pavement, and parks	Reduce city's consumption of gasoline, diesel, and natural gas Achieve at least 30% renewable energy use Reduce direct and indirect GHG emissions	Increase access for development of community gardens Improve access to farmer's markets Increase the number of sustainable redevelopment projects, and certified buildings	Increase on-street bike lanes to 70 miles Develop new sidewalks Decrease total vehicle miles traveled by city employees	All city employees involved in the National Incident Management System will maintain 100% of the training requirements to ensure preparedness Regionalize emergency preparedness planning

Summary

As risks and threats from climate change and global economy are factored into a decision-making process, communities start planning for resilience and sustainability. The ability to communicate and implement a long-term vision for the organization is instrumental for an effective sustainability and resilience strategy. Moreover, sustainability and resilience planning assists in the integration of all the elements of the Quadruple Bottom Line, including the often-overlooked components of governance. Local governments are confronted with many challenges, obstacles, and threats, including the effects of globalization, fiscal uncertainties, increased demands for services, and changing demographics. Contemporary organizations design their systems using effective sustainability and resilience planning to withstand external and internal pressures for maximum resiliency in dynamic environments, including climate change, and growing economic, environmental, governance, and societal pressures (Alibašić 2018a). Furthermore, the local governments deploy sustainability as organizational strategy (Alibašić 2018b).

The geopolitical, financial, international, and socio-demographic trends, the evolving nature of technology, and a fundamental shift in values generate challenges and opportunities for organizations. A transformative, sustainable, and resilient organization focuses on continuous improvement while encouraging and enabling the learning. As organizations evolve and transform, they continually reinvent themselves, invest in their continuing development, and evaluate assets while identifying core competencies. In using Plan-Do-Check-Act cycle in sustainability and resilience planning, organizations facilitate a learning environment and continual development with the goal of growing and changing and becoming more resilient in a dynamic environment. In identifying external pressures and moving forward, local governments need to evaluate the necessity of having both the strategic plan and sustainability plan. Adopting a sustainability and resilience plan and using it as a strategic plan and guidance document increases synergies in implementing initiatives and executing the vision, mission, and goals.

Further Discussions

- Discuss the steps involved in setting the stage for sustainability and resilience planning within an organization.
- Evaluate plans that cities have in place to address climate change and climate preparedness and readiness.
- Analyze the resources in place to implement sustainability and resilience planning.
- Assess the cities' programs of environmental protection, social justice and equity, economic opportunities, and governance.
- Define the dissimilarity between strategic planning and sustainability planning.
- Analyze the benefits, opportunities, risk, and limitations of the sustainability and resilience planning process.
- Further align sustainability and resilience planning within organizations using the concepts of Quadruple Bottom Line (QBL).

References

Alibašić H (2013). Grand rapids publishes climate resiliency report. TriplePundit people, planet, profit. www.triplepundit.com

Alibašić H (2017) Measuring the sustainability impact in local governments using the quadruple bottom line. Int J Sustainability Policy Practice 13(3):37–45

Alibašić H (2018a) Redesigning organizations for maximum resiliency in dynamic environments. In: Farazmand A (ed) Global encyclopedia of public administration, public policy, and governance. Zurich, Springer International Publishing AG, Cham. https://doi.org/10.1007/978-3-319-31816-5

Alibašić H (2018b) Sustainability as organizational strategy. In: Farazmand A (ed) Global encyclopedia of public administration, public policy, and governance. Springer International Publishing AG, Cham. https://doi.org/10.1007/978-3-319-31816-5_3433-1

Boswell MR, Greve AI, Seale TL (2012) Local climate action planning. Island Press, Washington, DC

City of Grand Rapids (2015) Fifth year sustainability plan progress report. Office of Energy and sustainability. [Alibašić H and Gosztyla D (eds)] Retrieved from: https://www.grandrapidsmi.gov/files/assets/public /

departments/office-of-sustainability/reports-and-documents/office-of-sustainability/5th-year-progress-report-sustainability-plan.pdf
City of Grand Rapids (2016) Sustainability plan FY2017-FY2021. Office of energy and sustainability. [Alibašić H (eds)] Retrieved from https://www.grandrapidsmi.gov/files/assets/public/departments/office-of-sustainability/reports-and-documents/office-of-sustainability/sustainability-plan-fy17-fy21.pdf
City of New Orleans (2009) City of New Orleans carbon footprint report. Retrieved from https://www.nola.gov/getattachment/7aeaa7e4-7489-48f0-ac7b-4406c0926e79/Appendix-Ch-13-City-of-New-Orleans-Carbon-Footprin/
City of New Orleans (2015) Resilient New Orleans: strategic actions to shape our future city. Retrieved from http://resilientnola.org/wp-content/uploads/2015/08/Resilient_New_Orleans_Strategy.pdf
City of New Orleans (2017) Master plan. Retrieved from https://www.nola.gov/city-planning/master-plan/
Coyle SJ (2011) Sustainable and resilient communities: a comprehensive action plan for towns, cities, and regions. Wiley, Hoboken
Cumo F, Garcia DA, Calcagnini L, Rosa F, Cumo F, Sferra AS (2012) Urban policies and sustainable energy management. Sustainable Cities Soc 4:29–34. https://doi.org/10.1016/j.scs.2012.03.003
Droege P (2006) The renewable city: Dawn of an urban revolution. Bull Sci Technol Soc 26:141–150. https://doi.org/10.1177/0270467606287531
Elkington J (1997) Cannibals with forks: the triple bottom line of the 21st century. Capstone Publishing Limited, Oxford, UK
Kennedy S, Sgouridis S (2011) Rigorous classification and carbon accounting principles for low and zero carbon cities. Energy Policy 39(9):5259–5268. https://doi.org/10.1016/j.enpol.2011.05.038
Lindfield M (2010) Cities: a global threat and a missed opportunity for climate change. Environ Urban ASIA 1(2):105–129. https://doi.org/10.1177/097542531000100202
Meridian Township (2017) Climate sustainability plan. Meeting our climate action and green community goals. Retrieved from http://www.meridian.mi.us/home/showdocument?id=5286
West Michigan Environmental Action Council (WMEAC) (2013) Grand rapids climate resiliency report. Retrieved from: https://thewmeacblog.files.wordpress.com/2013/12/grand-rapids-climate-resiliency-report-master-web.pdf

Identifying the Internal and External Stakeholders, the Level of Organizational and Community Engagement, and the Target Champions

3

"Cities are fantastically dynamic places, and this is strikingly true of their successful parts, which offer a fertile ground for the plans of thousands of people." Page 14, The Death and Life of Great American Cities, Jacobs, J. (1961). New York, NY: Vintage Books.

Key Questions
The third chapter of the book is directed at answering the following underlying assumptions and questions:

- How to identify stakeholders as part of the sustainability and resilience planning?
- What is the relevance of a stakeholder analysis and why is it so valuable to organizations?
- What is the relationship between identifying the right stakeholder and acceptance of sustainability and resilience planning?
- Who are the key stakeholders in ensuring successful sustainability and resilience campaign?
- How to engage a community and identify right partners for sustainability and resilience planning?
- What are the state and national partnership opportunities for implementing sustainability and resilience activities?

Introduction

The third book chapter covers an often-overlooked aspect of sustainability and resilience planning, stakeholder identification, and the level of staff and community engagement, including designs necessary to implementing the strategies and initiatives. Stakeholders for purposes of the sustainability and resilience strategic planning are referred to as internal and external stakeholders. Internal stakeholders include the elected and appointed officials, staff, and part-time employees. External stakeholders are residents, community members, partners, nonprofit and business organizations and their representatives, boards and task forces, vendors, contractors, and regional, state, and national organizations. A sequence of the appropriate steps in identifying internal and external stakeholders for organizational sustainability and resilience planning is explained. Similar to the process of mapping out the sustainability and resilience planning, stakeholder engagement process is multilayered, and it involves different stakeholders, internal and external. An overview of the synergies between departments, partners, leaders, and staff working to achieve sustainability within an organization is explained. Several subsections of stakeholder identification are described in-depth throughout the chapter.

The initial stage to sustainability and resilience planning encompasses identifying appropriate staff, defining roles and responsibilities, and naming target champions. The sustainability

H. Alibašić, *Sustainability and Resilience Planning for Local Governments*, Sustainable Development Goals Series, https://doi.org/10.1007/978-3-319-72568-0_3

and resilience planning process may be used as an opportunity to empower employees to own the designated targets, promote leadership and accountability, and further the collaboration between departments. Sustainability and resilience planning enables organizations to use a multifaceted, cross-sectoral approach for the improvement of operational efficiency. The primary drivers for successful sustainability and resilience programs are the requirements of communities and organizations to change and to adapt to the changes in the environmental, societal, economic, and governing conditions. The process starts with vision and mission statements, a scan of current conditions and available resources, existing sustainability efforts, and identifying stakeholders, both internally and externally. Most local governmental units have already sustainability efforts in existence through recycling, composting, energy efficiency, and biking programs. Stillmann III (1996) noted a frequent "wide gap between what is often promised by elected officials on the campaign trail and what can reasonably and effectively be delivered by the government" (p. 238). Embedding sustainability and resilience strategies within an organization will ensure continuity of such efforts regardless of the changes in leadership.

Sustainability and resilience planning is an instrument for multilevel stakeholder involvement to identify priorities for long-range goals and initiatives established. Additionally, sustainability and resilience planning provides appointed administrators and employees with a clear sense of ownership of the strategic goals and benchmarks for the expected results. The sense of responsibility for an organization's considerations creates a clear focus on results, the ability to adjust elements as necessary, and a sense of tenure to attain long-term goals. Through active stakeholder engagement, organization's vision is communicated internally and externally. New programs place additional pressure and duties on an existing workforce, and active stakeholder engagement reduces ambiguities over the actual burden of adding new programs and projects to the current staff.

Sustainability and resilience strategy ensures all staff, elected and appointed officials of local governments, and external stakeholders, including community residents, businesses, contractors, and vendors, are aware of the long-term strategic goals that an organization pursues. Doppelt (2010) noted the organizations with "the most progress toward sustainability understand that the shift to a circular economic model requires the full involvement of all of their internal members, as well as external stakeholders" (p. 36). Furthermore, having set resilience and sustainability steps in place, an organization demonstrates a commitment to reach those goals and establishes a vision for an improving and evolving organization. Once an organization's long-range goals, views, mission, and objectives are articulated, through its sustainability and resilience planning, organization is bound to obtain and attain specific results.

To determine community needs and to identify and implement appropriate sustainability and resilience solutions, the planning process encompasses participation from across the spectrum of sectors of the community. As each community faces its unique challenges and contains stakeholders, approaches need to be adjusted to fit those community features. Without recognizing the individual community needs and identifying the appropriate group of stakeholders, sustainability and resilience efforts may falter and ultimately fail. Sustainable and resilience strategy accounts for external socioeconomic pressures and the present conditions. Likewise, adequately aligned sustainability and resilience strategies ensure attention is paid to the most crucial and relevant topics to adhere to vision while combining and adjusting for socioeconomic concerns. Lastly, sustainable and resilience strategy will ensure that internal and external stakeholders are cognizant of the future direction of the organization.

Sustainability and Resilience Planning and Change Agents

Sustainability and resilience planning requires and demands a change in perspectives and approaches to organizational planning. When executing transformative changes with an ultimate goal of sustainability and resilience outcomes in organizations, the following represents indispensable elements in efficiently identifying internal stakeholders. The first step is to ensure that staff involved in writing a budget and implementing projects are engaged in evaluating sustainability efforts. These employees are change agents within an organization and the most effective advocates for sustainability and resilience planning. Besides, the top management, elected and appointed officials must be fully supportive to implementing changes and ensuring accomplishments.

Implementing change includes finding a starting point, attaching the benchmarks and measurements to existing efforts, constant communication, sharing progress information, involving staff and community, managing internal and external stakeholders, and working toward overall embedding of sustainability and resilience within organization. Contemporary leaders, managers, and community members are dependent on fast-paced information that can change on an hourly basis. The new leadership style and strategies deployed are necessary for organizations to survive and to continue to provide effective and efficient services. Effective leaders empower and enable employees to implement and efficiently execute sustainability and resilience strategies.

As argued by Yukl et al. (2010), "the ultimate source of organizational culture is the people who make up the organization" (p. 510). Moreover, Pfeffer (2010) was interested in "human sustainability is that developing a consistent set of measures or indicators of the construct, gathering data on them, and publicizing such data might provide more impetus for focusing on the human sustainability implications of what organizations do" (p. 41). In local governments, staff interact on a daily basis on a plethora of issues. Employees respond to concerned residents' complaints, political pressures, and delivery of service and thus represent the organization externally and internally. These employees are the face of the organization to the external customers. As sustainability and resilience measures in economic, social, and environmental, and governing areas become a part of the organizational dynamic, they facilitate teamwork and leadership and transformational changes.

Empowering Employees to Champion Sustainability and Resilience

The sustainability and resilience plan may serve as an empowerment mechanism for employees who own the identified targets. Staff collaboration leads to an improved collaboration and governance in organizations. The targets created with the help and input from employees are connected to the budget process. Target champions are designated to take ownership of specific target. Each champion of the sustainability and resilience target works closely with counterparts in other departments to ensure targets are being met. The creation of the plan is an organic process, not dictated by the mayor or the city manager but facilitated through the involvement of city departments at various management levels. Delineation of duties between multiple departments allows for removing strict silos.

Successful organizations encourage employees to seek more efficient ways to deliver a service and to be more proactive in finding solutions. Leaders try to empower employees by listening to their ideas and building on those. In the words of Pink (2010), "effective leaders are very good at hearing other people's stories and creating a story that allows people to see themselves as part of that narrative" (para 3). The main focus of sustainability and resilience leadership is to get to the sustainable future, empowering employees to carry out the vision for the entire organization. Such empowerment includes ethical behavior in the best interest of an organization where it becomes an ethical duty of employees to "challenge decisions that are misguided or unethical" (Yukl et al. 2010, p. 140).

The strength of the organization comes from within and is carried by its employees. When local government employees are encouraged to take more proactive, participatory, and mission-oriented role, organizations promote leadership and direct accountability. They are more likely to improve the service delivery. In return, such personal leadership attitude enables employees to accomplish objectives with fewer resources, to move forward on specific projects, and to be open to different points of view on solutions, to the point of being able to accept failure. In essence, the sustainability and resilience planning leads to a self-leadership development through empowerment. Such "strategic leadership practices can help" the organization "enhance performance while competing in turbulent and unpredictable environments" (Ireland and Hitt 1999, p. 45).

Leadership in Modern Organizations

Modern organizations are looking at ways to simultaneously engage employees vertically and horizontally to drive innovativeness as people are "inherently active and engaged, as long as they are treated fairly" regardless of position within an organization (Pink 2010, para. 4). In organizations focused on sustainability, innovativeness, and transformation, "followers are necessary for the successful performance of work" as much as the leaders are (Yukl et al. 2010, p. 139). Often, there is a gap between the vision and financial reality, and that gap is filled with employees carrying the heavy burden of balancing the duties and trying to be more proactive in their roles. Staff actively engaged in defining and crafting sustainable and resilience policies and implementing related programs are better equipped in managing them. Developing leadership and maintaining a reliable and loyal workforce are critical to the longevity of sustainability and resilience efforts. With funding from the NOAA Coastal Storms Program to disseminate climate change data and seek support for adaptation strategies to strengthen the community resilience of the city of Biloxi, administrators identified the "city's emergency manager, floodplain manager, engineer, and Community Rating System coordinator," to disseminate climate change data (City of Biloxi n.d.). Identifying the right staff and stakeholders is a crucial step to successful implementation of sustainability and resilience strategies.

Nutt (2002) contended "unmanaged social and political concerns, such as job security and vested interests, scuttle many persuasion-based implementation attempts" (p.98). It is imperative for the sustainability and resilience planning to include outcomes and targets related to improvements and training opportunities for employees to cope with an increased workload. For organizations in transition and need of constant changes, as suggested by Uhl-Bien et al. (2007), "adaptive leadership is an emergent, interactive dynamic that produces adaptive outcomes in a social system" (p. 306). Transformation and sustainability and resilience planning go hand in hand, and organizations pursuing both are interested in outcomes and are oriented toward performance. Tubbs and Schulz (2006) discussed "creating transformational change, developing an organizational culture that embraces continuous learning, building mechanisms to create and sustain change efforts, managing the change process, developing change agents, and encouraging individual as well as a structural change in the organization" (p. 32). These elements are critically important for nurturing leaders and empowering employees, where employees are critical stakeholders in implementing sustainability and resilience programs.

Adaptive leadership is employed in organizations where constant interaction changes occur regularly and continuous adaptations to economic realities are a must. Developing future leaders within a local government requires a strong commitment to diversity and various cultures. Often, challenging economic times drive local governments toward adaptive leadership, which can be used to foster sustainability, resilience, and overall transformation. Leadership in the local government can grow from within, and leaders must nurture the future leaders and prepare them for succession.

Assigning Target Champions

The sustainability and resilience plan sets specific goals with measurable targets and timelines for the achievement of these goals. The plan includes the assignment of responsibilities to various departments. For example, under the environmental outcomes and goals, one of the targets could be to increase the numbers of trees planted on private property by 10,000 by 2020. While the first champion assigned to this target could become an employee from the tree division of public works department, staff from other divisions such as parks and recreation and other departments are also involved in supporting the outcomes related to this particular target, leading to better collaboration and leveraging of resources and ultimately expediting progress. As city administrators encourage and nurture leadership and ownership within an organization, a similar strategy needs to be applied at a community level. A proactive process of community engagement, active surveys and other tools for seeking and hearing community input should be deployed. The targets created by employees and tied to the budget process are then championed and owned by those employees. Each champion of the sustainability target works closely with counterparts in other departments to make sure targets are being met.

As local government transition to sustainability and resilience planning, it helps to engage the employees through a planning process, which then they can own. Covey (2009) pointed out "when someone else's plan is imposed on us - it is a tough pill to swallow" (para. 4). Owning outcomes and processes provides for regular updates on progress and make recommendations relevant to specific services. Leaders recognize the need to get a buy-in from employees for successful transitions. Each team would include coaches responsible for the successful growth of an associate within a team. By actively participating in leadership development and leadership creation, the city can avoid the underlying tensions. As pointed out by Friedman (2004), "people tend to misperceive dynamic feedback in an organization" (p. 111). In assigning the champions for each target in the sustainability and resilience planning, organizations ensure all staff own and are accountable for the attainment of targets attached to them.

To create a sustainable and resilient community, every staff is accountable for her/his decisions and actions because it will impact others consequently. The target champions are responsible for monitoring the progress of the assigned targets but also have a supporting role toward the attainment and measurement of targets. A sample process of assigning target champions and delineating direct and indirect functions is provided in Table 3.1 entitled Target Champions. While the table represents a sample of the city's sustainability plan, layered in the following sequence, theme > goal > outcome > targets, it shows the relevance in identifying direct and indirect champions of sustainability and resilience targets. It is a based on a tentative template and is meant to be used to share with staff for additional input, changes to the proposed targets, and identification of appropriate stakeholders (staff).

Community Engagement

Community engagement is an indispensable facet of sustainability and resilience planning for local governments or any other organization. The informed community understanding how the sustainability and resilience planning improves service delivery and how it advances organizational efficiency is more supportive of sustainability programs and is proactively engaged and even support it financially. The key to successful community engagement is to identify the uniqueness of the local government and the community and to be able to apply strategies to fit those characteristic. Community's participation, partnerships, stakeholder engagement, and the commitment to the pursuit of sustainability and resilience are essential to successful sustainability and resilience planning.

As cities transition their planning toward sustainability and resilience, they work on getting a backing from the community using an innovative process of engaging residents and business owners through transformation advisory groups, and sustainability task forces. Mascarenhas et al. (2015) argued for the inclusion of a broad range of perspectives to be represented by various stakeholders when planning for sustainability.

Table 3.1 Identifying target champions in a Quadruple Bottom Line sustainability and resilience plan

Target #	Target wording	Outcome champion	Comments/metrics/support/ revisions/research
Theme 1: Economic growth and opportunities			
Goal 1: Create a nurturing environment for buisnesses			
Outcome 1.1: Adopt innovative, entrepreneur-focused economic development strategies that leverage the resources of the city to maintain the economic vitality of the community as a whole			
1.1.1.1	Increase the number of new businesses locating in the city with the assistance of the economic development staff by # between DATE and YEAR	Director/ assistant	Economic development with chamber of Commerce and other economic development agencies to support this target
Goal 2: Facilitate job creation and job growth			
Outcome 2.1: Facilitate sustainable business development to support job creation using tax incentives and other available economic development tools			
1.2.1.1	--% of jobs created or retained with incentives will be permanent, full time employment with benefits annually	Director	Economic development and local business agencies to collaborate on this outcome. Seek support from HR and other city departments
1.2.1.2	Increase the number of micro local business Enterprise (micro-LBE) contained in the City directory by --% over FY15 results by DATE	Director	Economic development and local business agencies to collaborate on this outcome. Seek support from HR and other city departments
Theme 2: Resilient and safe neighborhoods			
Goal 1: Promote quality design and construction			
Outcome 1: Promote quality design and construction consistent with neighborhood character which encourages efficient land use, green building design, and walkability			
2.1.1.1	Decrease the number of vacant lots or brownfields by --% per neighborhood, while increasing by --% the number of completed lots that Preserve historic buildings and repurpose vacant lots and brownfields by DATE, YEAR	Director/ assistant	Metric: % of brownfields per neighborhood & % of projects that preserve historic buildings Alternative metrics: % of land designated by city to preserve open space % of land designated as historic preservation
Goal 2: Vital neighborhood infrastructure			
Outcome 2: Promote quality neighborhood infrastructure (condition of housing, city streets, and sidewalks; availability of parking; existence of trees and green space; and access to parks and recreation amenities) as an important element of any neighborhood			
2.2.2.1	Increase the number of affordable housing units by # by DATE		Community development/ econ. Dev.
Theme 3: Resilient environment			
Goal 1: Reduction of energy footprint			
Outcome 1: Implement initiatives to counteract the effects of greenhouse gas emissions (GHG) to provide a cleaner and greener community and a higher quality of life			

(continued)

Table 3.1 (continued)

3.1.1.1	Reduce the city's greenhouse gas (GHG) emissions to % percent below YEAR levels by YEAR	Energy	Inventory- FROM YEAR WITH A total CO2e was estimated to be # OF metric tons. With this TARGET, aiming to reduce levels to # metric tons CO2e An alternative TARGET could be --%.... # metric tons CO2e
3.1.1.2	Achieve ---% of energy use from renewable sources such as hydro, wind, solar, and geothermal by DATE, YEAR	Sustainability, environmental services	
Goal 2: Strengthening of climate protection and resiliency			
Outcome 1: Integrate operations and preparedness measures into EXISITING plans to respond to climate change related threats and disasters			
3.2.1.1	Acknowledge climate change in infrastructure planning by considering best available climate projections and incorporating climate adaptation planning into capital, operating, and maintenance programs by DATE, YEAR.	Sustainability, other departments	
Goal 3: Waste minimization and expansion of re-use and recycling opportunities			
Outcome 1: Expand reuse and recycling opportunities, composting of yard waste, to decrease the amount of waste sent to landfills			
3.3.1.1	Reduce the amount of landfill contributions by --% within # of years	Director	Public works in partnership with county
Goal 4: Protection and enhancement of natural systems			
Outcome 1: Integrate protection and restoration of natural systems into plans as they provide ecological "services"			
3.4.1.1	Increase tree canopy coverage by --% per neighborhood by DATE, YEAR.	Director	D parks and recreation
Theme 4: Good governance			
Goal 1: Accountable, accessible, and transparent service delivery			
Outcome 1: Strengthen financial management, and reduce operational costs			
4.1.1.1	Increase cost avoidance due to energy inefficiency by an additional --% over FY-YEAR results in city facilities by DATE	Energy staff	All energy user departments to coordinate
Goal 2: Effective engagement of stakeholders			
Outcome 1: Communicate decision-making process outcomes in a clear and understandable manner			
4.2.1.1	Ensure that over # OF hours per week of government access Cable Channel is diverse and additional language programming, annually	Analyst II	Work with translation agencies to translate programming
Goal 3: Create an open and inclusive government			
Outcome 1: Ensure services are easily accessible to a diverse customer base through proven best practices and coordination across all service channels			
4.3.1.1	Increase the use of online permitting by an additional % over FY--results by DATE	Planner	
Goal 4: Provide effective and efficient service delivery			
Outcome 1: Implement decisions and follow processes that make the best use of resources to serve the needs of the entire community while balancing competing interests			

(continued)

Table 3.1 (continued)

4.4.1.1	Respond to % of street lighting outages within # OF hours of being reported annually	Director	Street light/public works
4.4.2.2	Increase the city's overall fire code inspection completion rate to % by DATE	Fire chief	Inspections

Adapted from City of Grand Rapids Sustainability Plan

Depth and breadth of identified stakeholders will increase the likelihood of a successful implementation of sustainability- and resilience-related efforts. However, the mere inclusion of community members in the planning stages of sustainability and resilience planning is not enough for success. The sustainability and resilience achievements in the community are dependent on the collaborative efforts of all local government departments dealing with energy, waste minimization, business development, nonprofit organizations, government agencies, human resources, diversity, community and environmental stakeholders, and the citizens. Hitchcock and Willard (2008) proposed "sustainability solutions often require an interdisciplinary, multi-stakeholder approach, involving people from across the organization or even from multiple organizations" (p. 119).

As cities begin their sustainability and resilient planning process, it is critical to start building these collaborative relationships early to complement the support from city administrators, elected and appointed officials. Sustainability action task force, teams, and cooperative groups are beneficial to the sustainability and resilience efforts. Strategies for engaging community members are endless. For example, in response to community-wide demands for sustainability, the city of Beaverton, OR, hosted a series of community engagement forums to discuss sustainability efforts within Beaverton. Furthermore, the city led the Revitalization Roundtable, considering the partnership opportunities with the regional, state, and federal agencies and stakeholders, to among many other efforts encourage sustainability and build sustainable community (NLC 2013). The roundtable was one of many examples of the city's concerted effort for a meaningful and organic community engagement, seeking input and providing opportunities for various stakeholders to contribute toward sustainability of the community.

Portney and Berry (2010) discussed how sustainable communities have a greater participation in processes by its residents and other interest groups. Sustainability and resilience, in this case, may be viewed using the lens of interest groups and participation building. The authors attempted to answer the "empirical questions about participation" as an essential aspect and pursuit of sustainability in the context of civic engagement in urban cities of the United States (Portney and Berry 2010, p.120). Communities with sustainable and resilient policies in place have a participatory citizenry, aspects of social and governance factors, vital in viewing sustainability and resilience in cities as an avenue for more citizen engagement and input.

Resilient Local Government Spotlight: Community Engagement for Resilient DC

The Washington DC's Office of Resilience was set up to deal with "both catastrophic shocks and chronic stresses in order to ensure that DC thrives in the face of change," including "the ability to withstand any natural or man-made challenges that threaten our communities and tackle the social challenges that come with being a fast-growing city" (Office of Resilience 2017). Some of the significant stressors and shocks identified by the DC Office of Resilience are threats of potential terrorist attacks, road congestions, and climate change-induced extreme weather, including heat waves. The DC's resilience office was set up with financial and technical support from 100 Resilient Cities (100 RC). The DC Office of Resilience began an extensive process of community engagement, including surveys to identify strengths and weaknesses in the community.

Brugmann (2012) discussed scaling climate adaptation strategies in urban settings, describing resilience as "the ability of an urban asset, location and/or system to provide predictable performance" (p. 217). Cities' unique role is to provide services and identify policy options in the best interest of the organization and community at large. Seeking resident and business input, through active community engagement, is a vital component for a resilient and sustainable community. The Office of Resilience in DC has an ultimate goal of developing a "resilience strategy [that] will be a holistic, action-oriented plan to build partnerships and alliances as well as financing mechanisms, and will pay particular attention to meeting the needs of vulnerable populations" (Office of Resilience 2017).

Partnership Opportunities

In identifying internal stakeholders and implementers, and external supporters, local governments map out sustainability and resilience planning process, and feature the auxiliary effect on a community. As part of the process of community engagement, municipalities seek partnership opportunities and develop a culture of continuity and adaptability in planning. While the local governments' sustainability and resilience planning relates to an organizational operations and resources, it also focuses on the broader implication of sustainability and resiliency in the region and impact on the community. Collaborative resilience and sustainability planning is accomplished through variety of measures and initiatives, including the energy audits, efficiency improvements in residential homes, and promotion of recycling through local businesses.

Resilient City Spotlight: City of Cincinnati

In 2013, local government in Cincinnati developed the Green Cincinnati Plan, with strict focus on sustainability. Through a series of four public meetings, the Office of Environmental Quality in Cincinnati engaged more than 200 members of the Green Umbrella Action Teams and other community members to assist and provide input and recommendations for the Green Cincinnati Plan. The Green Umbrella is a regional sustainability alliance in the region around Cincinnati. In addition, a steering committee consisting of representatives from both the public and private sectors provided leadership in reviewing and finalizing the recommendations from the Green Umbrella Action Teams. The steering committee made the final recommendations to the city administrators after holding a charrette (City of Cincinnati 2013).

Practical Application: Partnership for Sustainability and Resilience

In Grand Rapids, Michigan, collaborative efforts between the public and private sector organizations aimed at promoting and sharing emergent sustainability practices have been in place since 2004. In partnership with the local government officials, local colleges and universities, private sector companies, and non-profit organizations are seeking the best solutions to region-wide issues. Some of the regional sustainability and resilience-related collaborative partnership activities included the writing of the Grand Rapids Climate Resiliency Report, collaboratively competing for the federal resilience grants in partnership with the state and completing a river restoration planning, coupled with floodwall improvements.

Regional, National, and International Partnership Opportunities

Working through regional and national organizations is a beneficial form of partnership for delivering and implementing sustainability- and resilience-related initiatives. Cities work with ICLEI – Local Governments for Sustainability, US Conference of Mayors (USCM), National League of Cities (NLC), and other national-level organizations involved in sustainability. However, cities also partner and share best practices through regional partnership opportunities. For example, over hundred cities, members of the Great Lakes-Saint Lawrence Cities Initiative (GLSLCI), promote a multi-state, bi-national strategy to curb environmental and climate change threats to the most significant body of fresh water on Earth, including Asian carp and other invasive species threatening the Great Lakes ecosystem. Beyond water quality advocacy in addressing nutrients and algae bloom, biodiversity, micro-plastics, oil transport, and nuclear energy, the organization undertook a myriad of other initiatives and actions to support the members' programs for climate adaptation and mitigation in the Great Lakes and Saint Lawrence water basin (GLSLCI 2017). In DC, the GLSLCI advocates for the Great Lakes Compact, seeking funds and plans for the Asian carp threats to the Great Lakes and advocating for separation of the Great Lakes and Mississippi River watersheds to safeguard from aquatic invasive species through the Great Lakes and St. Lawrence Cities Initiative. Collaborative initiatives in the areas of sustainability and resilience planning in the region, energy audits, energy efficiency improvements in neighborhood homes, increased recycling through local economic development incentives, and the regional climate action reporting are further evidence of the importance of partnerships to achieve prosperous sustainability and resilience-related outcomes. In the words of the current chair of the board of directors, Mayor Paul Dykstra of Niagara Falls, NY, "by using their integrated approach to environmental issues," the GLSLCI member mayors from the United States and Canada representing their respective cities "form a force to ensure the long-term sustainability of these precious resources for future generations" (GLSLCI 2017).

Summary

Successful integration of sustainable and resilience practices requires assessment and sensitivity related to the culture of an organization, the professions within that organization, and the demographics of key stakeholders. The universal language of sustainability and resilience planning should focus on well-known principles but be spoken in terms that resonate with key stakeholders as a mechanism to engage them in a shared vision. The formalized process of sustainability and resilience planning is critical. Moreover, "cities' resilience efforts are flexible, with a firm leadership commitment and steady comprehension of the new realities on the ground" (Alibašić 2018a, p.4). Finally, the commitment to sustainability and resilience planning leads to improved outcomes (Alibašić 2018b). Sustainability and resilience planning process encompasses social, economic, environmental, and governance objectives to improve operational efficiency and community resilience and provides a framework for meeting goals and objectives. According to the National Centers for Environmental Information (NCEI), the US communities experienced 219 weather and climate disasters with the total cost in damages as a result of these 219 events exceeding $1.5 trillion, including "the initial cost estimates for Hurricanes Harvey, Irma and Maria" (NOAA 2018). These climate-related and recurring extreme weather events put an extraordinary strain on communities and local governments tasked not only with disaster response but also with the post-disaster recovery. The resilience planning with adaptation, mitigation, and disaster preparedness elements are paramount for local government organizations.

By expertly identifying and engaging key stakeholders, cities seek acceptance and support for essential community's sustainability and resilience initiatives. Moreover, sustainability and resilience planning includes a process of defining the strategy; expressing vision, goals, and objectives; and communicating those to its internal and external stakeholders. The sense of ownership over an organization's goals provides both internal and external stakeholders with a necessary comprehension of the sustainability and resilience plan's goals and objectives. Additionally, a well-drafted strategic resilience and sustainability plan expresses a vision while empowering stakeholders, namely, employees to exceed expectations in implementing initiatives. With a clear vision, strategic sustainability and resilience plans establish the course for an organization's future.

Further Discussions

- Analyze the importance of gaining a buy-in by internal and external stakeholders to sustainability and resilience planning.
- Explain the points of understanding and analyzing stakeholders for resilience and sustainability planning.
- Discuss how organizational leadership can avoid pitfalls in stakeholders' engagement.
- Create a stakeholder assessment of external and internal actors in sustainability and resilience planning process.
- Assess the connection of stakeholder analysis with that of environmental and resources scan.

References

Alibašić H (2018a) Leading climate change at the local government level. In: Farazmand A (ed) Global encyclopedia of public administration, public policy, and governance. Springer International Publishing, Cham. https://link.springer.com/referenceworkentry/10.1007%2F978-3-319-31816-5_3428-1

Alibašić H (2018b) Ethics and sustainability in local government. In: Farazmand A (ed) Global encyclopedia of public administration, public policy, and governance. Springer International Publishing, Cham. https://doi.org/10.1007/978-3-319-31816-5_3427-1

Brugmann J (2012) Financing the resilient city. Environ Urban 24:215–232. https://doi.org/10.1177/0956247812437130

City of Biloxi (n.d.) NOAA coastal storms program. Retrieved from http://www.gulfcoastplan.org/wp-content/uploads/2014/07/1404749266-fact-sheet-biloxi_final-2.pdf

City of Cincinnati (2013) Green Cincinnati plan. Retrieved from https://www.cincinnati-oh.gov/oes/linkservid/6CE53223-9206-9F36-DB7FA3444F16A1A0/showMeta/0/

Covey SMR (2009) How the best leaders build trust. Leadership now. Retrieved from: http://www.leadershipnow.com/CoveyOnTrust.html

Doppelt B (2010) Leading change toward sustainability. Greenleaf Publishing Limited, Sheffield

Friedman S (2004) Learning to make more effective decisions: changing beliefs as a prelude to action. Learn Organ 11(2):110–128

Great Lakes Saint Lawrence Cities Initiative (GLSLCI) (n.d.) Initiatives. https://glslcities.org/initiatives/past-initiatives/

Hitchcock DE, Willard ML (2008) The step-by-step guide to sustainability planning: how to create and implement sustainability plans in any business or organization. Earthscan, London

Ireland D, Hitt M (1999) Achieving and maintaining strategic competitiveness in the 21st century: The role of strategic leadership. Acad Manage Exec 13(1):43–57. Retrieved from Business Source Premier database

Jacobs J (1961) The death and life of great American cities. Vintage Books, New York

Mascarenhas A, Nunes LM, Ramos TB (2015) Selection of sustainability indicators for planning: combining stakeholders' participation and data reduction techniques. J Clean Prod 92:295–307

National League of Cities (NLC) (2013). Revitalization roundtable kicks off local- federal conversations on sustainable manufacturing, job creation and redevelopment in Beaverton, Oregon. Retrieved from http://www.nlc.org/article/revitalization-roundtable-kicks-off-local-federal-conversations-on-sustainable

National Oceanic and Atmospheric Administration (NOAA). National Centers for Environmental Information (2018) Billion-dollar weather and climate disasters: overview. Retrieved from https://www.ncdc.noaa.gov/billions/

Nutt PC (2002) Why decisions fail: avoiding the blunders and traps that lead to debacles. Berrett-Koehler Publishers Inc, San Francisco

Office of Resilience (2017) Resilient Washington DC: About resilient DC. Retrieved from https://resilient.dc.gov/page/about-resilient-dc

Pfeffer J (2010) Building sustainable organizations: The human factor. Acad Manage Perspect 24(1):34–45, 12p; Retrieved from Business Source Premier database

Pink D (2010) Analytics are out, autonomy is in. Walden Alumni Mag 5(2): 27–28. Summer/Fall 2010. Walden University. https://www.waldenu.edu/-/media/Walden/files/newsroom/alumni-magazines/walden-magazine-5-2.pdf

Portney KE, Berry JM (2010) Participation and the pursuit of sustainability in U.S. cities. Urban Aff Rev 46(1):119–139. https://doi.org/10.1177/1078087410366122

Stillman R II (1996) The American bureaucracy. The core of modern government, 2nd edn. Nelson-Hall, Inc. Chicago, IL

Tubbs S, Schulz E (2006) Exploring a taxonomy of global leadership competencies and metacompetancies. J Am Acad Bus Camb 8(2):29–34

Uhl-Bien M, Marion R, McKelvey B (2007) Complexity leadership theory: shifting leadership from the industrial age to the knowledge era. Leadersh Q 18(4):298–318

Yukl G, George JM, Jones GR (2010) Leadership: building sustainable organizations (Laureate education, Inc., custom ed.). Custom Publishing, New York

4 Measuring, Tracking, Monitoring, and Reporting Sustainability and Resilience Progress

"Proper internal control: The third and last necessity to achieve success in business is to have a system of internal control. With a system of internal control, business transactions are recorded in such a systematic way that one may understand each one of them at glance." Fra Luca Pacioli (1494) Particularis de Computis et Scripturis pp.1–2

Key Questions
The fourth chapter of the book is aimed at answering the following underlying assumptions and questions:

- Why do organizations measure sustainability and resilience outcomes?
- What do organizations measure to become more sustainable and increase resilience?
- What are some objective sustainability and resilience measurements?
- What targets are relevant to organizational sustainability and resilience planning?
- What is the difference between the indicators and targets?
- What are the differences and similarities in measuring and reporting the sustainability and resilience efforts in various communities?
- What type of reports are the most effective ways of communicating the progress of sustainability and resilience-related initiatives?
- How do local governments ensure transparency and accountability through their sustainability and resilience progress reports?

Introduction

The fourth chapter of the book examines the role and interests related to the measurement of sustainability and resilience efforts, activities, and initiatives. The foundation for sustainability and resilience planning is the ability of staff in organizations to measure, track, monitor, and report the sustainability and resilience-related outcomes. Chapter 4 covers measuring, tracking, monitoring, and reporting using the Quadruple Bottom Line strategy and developing a sustainability and resilience progress report to ensure accountability, transparency, and good governance. In guaranteeing successful implementation of sustainability and resilience, organizations analyze, evaluate, and disseminate information about the outcomes and goals and track and measure them in real time. The most appropriate and effective ways of tracking, monitoring, and reporting the sustainability and resilience efforts are assessed and analyzed, including an evaluation of the sustainability and resilience goals, objectives, and targets.

A proactive approach to measuring sustainability and resilience ensures the departments and city staff are being held accountable, with an effective ownership of the targets. Progress reports available online offer an additional answerability for the resilience and sustainability-related efforts. In balancing sustainability and resilience objectives, local governments report progress on each target annually. Measuring sustainability and resilience activities is non-negotiable. Quantifying and reporting results are critical to ensuring accountability and

H. Alibašić, *Sustainability and Resilience Planning for Local Governments*, Sustainable Development Goals Series, https://doi.org/10.1007/978-3-319-72568-0_4

transparency. The design of the sustainability and resilience programs does not necessarily lead to a more efficacy unless the programs and initiatives are effectively tracked, measured, and reported.

An example of efficient use of measuring the sustainability and resilience outcomes is the local governments' commitment to reducing greenhouse gas (GHG) emissions, often referred to as a carbon footprint reduction. Local governments express the commitments for carbon reduction through targets in a sustainability and resilience plan and monitoring and publicizing of such actions. Administrators measure outcomes related to GHG emission reduction for the entire organization. Benefits from reduced carbon footprint are interpreted as benefits for the community, as either direct savings or cost avoidance. A record of measurable outcomes contributing to a significant reduction of GHG emissions includes the waste minimization, energy improvements, renewable energy, electric vehicles, charging stations infrastructure, increased recycling availability, low-impact development, the addition of bike lanes, and aggressive tree planting programs. Local government administrators can convert the evidence of ecological advantage from carbon footprint reduction in operations into the gains for the entire community.

Organizations committed financially and otherwise to the sustainability and resilience goals and practical application and implementation of those efforts feature their progress prominently toward ultimate sustainability and resilience outcomes. By connecting sustainability and resilience planning directly to their budget process, organizations indicate and substantiate commitment to sustainability and resilience. Local governments track, measure, and report results. By measuring, tracking, and reporting data, staff can identify opportunities for cost-saving measures. For example, cities may be tracking and reporting the waste minimization or energy efficiency initiatives. Through a detailed measurement, administrators are able to gain a comprehensive understanding of the greenhouse gas emissions of its facilities and fleet, using the CO2 equivalent, as well as the emissions generated in the community from residential, commercial, industrial, and transportation-related activities. The resources and templates to feature and showcase progress are available to local governments seeking them.

Resilient County Spotlight: Resilient and Resource-Efficient Alachua County, FL

One of the early adopters of the term resilience in planning, Alachua County leaders, issued the declaration striving for the county operations to be more resilient and resource efficient. A more specific set of resilience quantitative and qualitative goals include the desire to reduce the community-wide use of liquid fuels by 2020, including:

- Increasing the vehicle occupancy and ridership by at least 25%.
- Doubling the fuel consumption efficiency of Alachua County government fleet.
- Moving to non-fossil fuel fleets as soon as possible. Encourage commercial and private fleets to accomplish the same.
- Reducing by 1/2 of the 13,500 annual (2008) miles driven by each registered Alachua County driver.
- Decreasing the single occupancy vehicle trips by 25%.
- Maximizing the mobility opportunities; creating infrastructure and pathways for electric vehicles; expanding public transit, telecommuting, and flexible operating hours; using carpools, ride sharing, and car share companies; and encourage the Metropolitan Transportation Planning Organization to adopt these policies and goals (Alachua County 2008, p.1).

The county does not provide an annual update or progress in meeting the specific goals in the declaration. However, as an early appeal to a countywide resilience, it is a remarkable document. An added feature would be a potential measurement of resilience. The next logical step for the county is to engage internal and external stakeholders, revisit the declaration, and start providing consistent progress updates and reports.

Measuring, Monitoring, Tracking, and Reporting (MMTR)

The foundation for sustainability and resilience planning is the ability of staff in organizations to measure, monitor, track, and report the outcomes. While some sustainability and resilience plans are outcome-based, a greater emphasis is assigned to measurable targets and results. A comprehensive sustainability and resilience planning in most cases can serve as a guideline to aid leaders and administrators in decision-making. By measuring or quantifying the stated goals in the sustainability and resilience plans, local government administrators have a better appreciation of budgetary and financial resources in avoiding the impending effects of recessions and downturns in the economy. Measurement of sustainability and resilience is vital for a successful functioning of an organization and improved performance. In addition to measuring economic benefits, measuring environmental, social, and governance effects of sustainability and resilience activities provides an advantage to comprehending the system's broad impact from local government's actions.

Local governments have an obligation to track, measure, and report results. By measuring, tracking, and reporting data, staff can identify opportunities for cost-saving measures and cost avoidance, seek innovative solutions to problems, and deal with potential interruptions to operations by reducing ambiguity and failures. Comprehensive knowledge of costs related to all facets of facilities and fleet, equipment, and processes, in general, is instrumental. Local governments may use the cost expressed in dollar or other currencies, CO2 equivalent, and may measure all the activities, including the greenhouse emissions generated in the community from residential, commercial, industrial, and transportation-related activities.

For example, Kennedy and Sgouridis (2011) sought to define a framework for measuring the greenhouse gas emission impact from cities, labeling it as a carbon accounting "adapted to the urban scale" (p. 5260). Measuring the greenhouse gas emission impact is often associated with the local governments' efforts to reduce a negative effect from its activities by using energy efficiency and renewable energy processes. Furthermore, authors posited that "clarifications regarding the system boundaries and the emissions scope for a particular city are essential to developing a strategy for monitoring and managing urban-level carbon emissions" (Kennedy and Sgouridis 2011, p. 5268). In essence, a correct accounting procedure for greenhouse gas emission inventory includes the precisely defined system boundaries. The flowchart entitled measuring, monitoring, tracking, and reporting sustainability effort provides a visual representation of the several facets and characteristics of an operative reporting mechanism of the local governments' sustainability and, by extension, the resilience activities and initiatives.

There is an explicit association between the ability to measure sustainability, resilience, and good governance. Hemmati and Enayati (2002) described good governance as "a core concept" and one that "is indispensable for building peaceful, prosperous, and democratic societies," demanding "consent and participation" (p. 41). Good governance and answerability create a positive return on investment, as a result of the greater acceptance of the resilience and sustainability practices within the organization and the community. The themes of answerability, good governance, and accountability succinctly link to reporting and measuring sustainability and resilience.

Key Features and Elements of Measuring, Mentoring, Tracking, and Reporting (MMTR)

1. Establishing goals, vision, mission, and targets
2. Determining benchmark data, year, and end year
3. Assigning target champions
4. Aligning the targets with budget/fiscal plans
5. Mentoring and tracking
6. Measuring using qualitative and quantitative data
7. Reporting and disseminating results and progress update (Fig. 4.1)

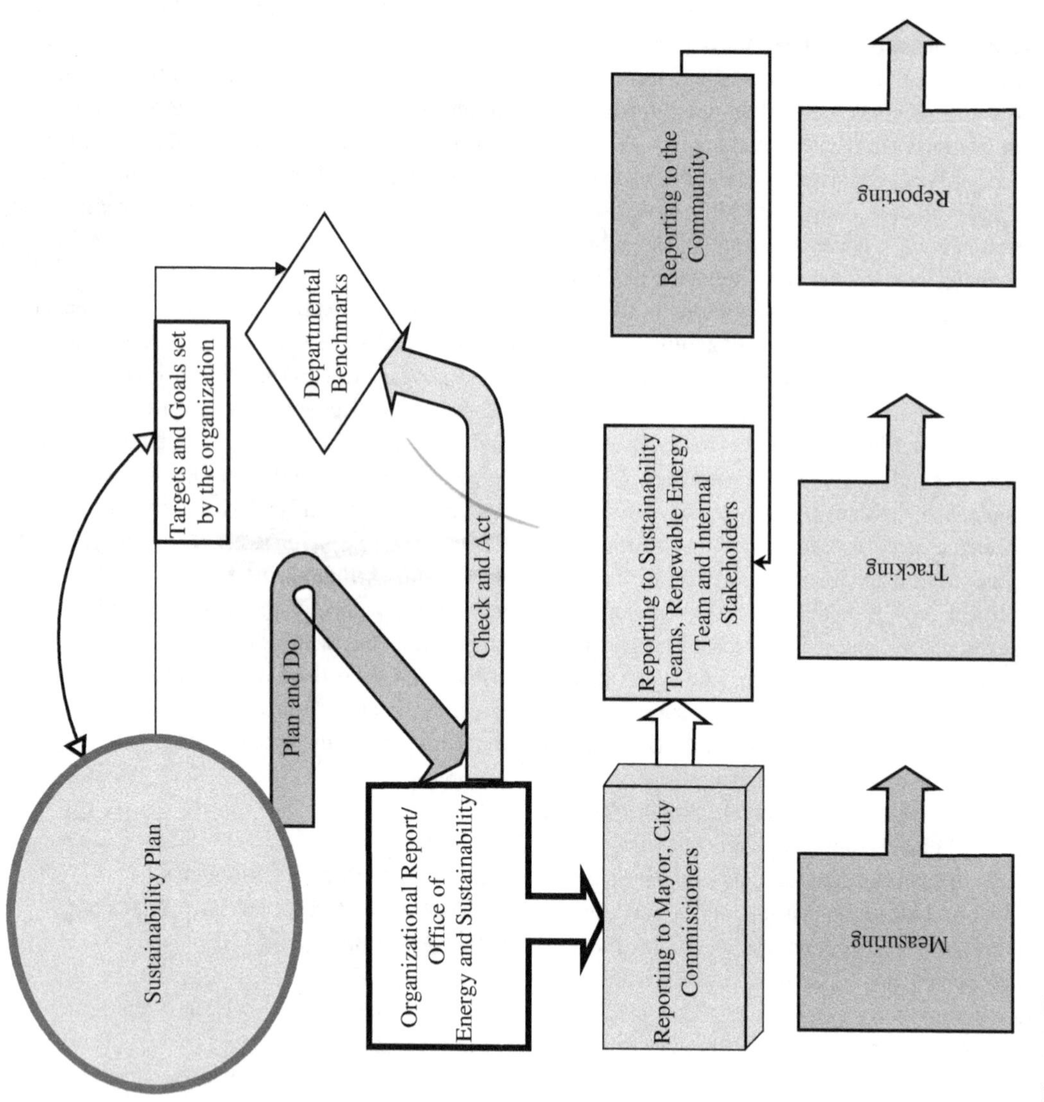

Fig. 4.1 Flowchart measuring sustainability and resilience

Why Do Cities Measure Sustainability and Resilience?

As a result of their direct and indirect activities, organizations cause the environmental, economic, social, and governing community-wide impact. Local governments measure an indirect and direct carbon footprint and economic and social dimension of their activities for better understanding, improvements, and corrective measures as necessary. As an imperative, the local governments measure their impact using benchmarks, baseline data, and targets and in alignment with their budgetary goals and to keep the stakeholders and constituents informed. Local governments' role is to deliver services and quality of life outcomes. Conventionally, municipalities are most concerned with their annual budgeting and service delivery and less with influencing and impacting the economic, social, environmental, and sound governance goals. Over 500 years ago, in describing how to keep the accurate inventory record and the daybook, Fra Luca Pacioli recommended no details be omitted from the reports, adding "the questions 'who, what, why, how, when and where' need to be answered" (Pacioli 1494, p. 11). Similarly, it is in the local governments' interest to keep an accurate and detailed record of their sustainability and resilience programs and activities.

There is a demand for sustainability and resilience outcomes at local government level and in the society in general. Elkington (2012) offered a five-stage model as the critical determinant for identifying sustainability measurements, "the 5Es that maps pathways to scale in the Zero Footprint Economy," including the "population growth, pandemics, poverty, pollution, and the proliferation of weapons of mass destruction" (p. 13). Bell and Morse (2008) argued against measuring sustainability as they put it "sustainability is not a 'thing' that can be measured, and an element of circularity appears inevitable: sustainability becomes defined by the parameters that can be measured rather than the other way around" (p. xvii). However, through measuring sustainability-related efforts, an organization proficiently conveys its characterization of sustainability, operationalizes it, and embeds into its operational framework.

In operationalizing sustainability and resilience metrics, a robust method for generating feedback and information sharing is developed. Metric, through internal and external networks, collaboratively developed and disseminated on a regular basis in a way that is understandable to all the stakeholders. Organizations measure sustainability and resilience as a way to show accountability and transparency as well as a way to validate vision statements, goals, and objectives. Local governments utilize qualitative and quantitative measurements in organizational systems as a way to seek feedback and preemptively avoid disruptions in operations. Based on a clear vision and outcomes, quantitative data can be used to monitor and measure capacity and delivery of results and any aspect of a service delivery. Qualitative data can be used to document the "who, why and how" of decision-making (Doppelt 2010). Quantitative and qualitative data are employed to evaluate the what of the administrative process. Measurements examine the communities' commitment to sustainability and resilience programs.

Quadruple Bottom Line Measurements

The most common approach to measuring sustainability and resilience activities in local governments is the concept of Triple Bottom Line (TBL), defined through economic, social, and environmental pillars. In general, it is impossible to place each target into a single silo, as most targets belong to more than a single category. Often, the structurally essential elements of sustainability planning such as transparency, accountability, community engagements, and good governance fall by the wayside and are not being measured by local governments. Conclusively, Alibašić (2017) described the sustainability planning by focusing on the Quadruple Bottom Line (QBL) and considerations of the capacity of organizations "to embed and incorporate a set of definitive policies and programs to address economic, social, environmental, and governance aspects of sustainability" (p. 41). Furthermore, governance contains the components of fiscal responsibility, participation, community engagement, transparency, answerability, and accountability (Alibašić 2017, 2018).

Resilience Spotlight: Sustainability Plan, Report, and Quadruple Bottom Line

In 2016, City of Grand Rapids, MI adopted the first of its kind of sustainability plan using the Quadruple Bottom Line for tracking sustainability targets. Local government administrators built in the Quadruple Bottom Line approach based on the previous iteration of the sustainability plan measuring and reporting progress of over 200 targets grouped under the themes of economic prosperity, social equity, and environmental quality. In the FY2017–2021 Sustainability Plan, the city staff added the fourth component under theme governance (Fig. 4.2).

By adding governance, the city sought to further good governance and answerability, fostering accountability and transparency in its operation. The new sustainability plan also drew from other plans the city had in place to align them with goals, outcomes, and themes. The Quadruple Bottom Line provided the overarching pillars: economic, social, environmental, and governance. The specific ten themes of this plan are housed under one of the four QBL pillars, and specific targets are categorized under separate goals and outcomes, directly connected to other city's plans:

- Economic opportunity
- Great neighborhoods
- Social equity
- Safe community
- Resilient systems
- Balanced transportation
- Sustainable assets
- Fiscal resiliency
- Transparency and accessibility
- Good government (City of Grand Rapids 2016, p. 3)

The advantage of the Quadruple Bottom Line sustainability planning using targets is in the well-defined objectives and goals, benchmarking with baseline years and reporting of the results. Moreover, in identifying and assigning target champions, ownership of the sustainability plan is guaranteed. Furthermore, all the components of the city's climate change, adaptation, mitigation, and resilience planning are included in the sustainability plan, where the sustainability plan is, in essence, a resilience plan and vice versa. Qualitative and quantitative target spotlights from the city's sustainability plan include the following:

1. Create incentives for development of socially responsible B Corporations by June 30, 2021.
2. Increase the diversity of neighborhood business districts by supporting the establishment of 20 new businesses annually.
3. Ensure that 80% of jobs created or retained with incentives will be permanent, full-time employment with benefits annually.
4. Increase the private business investment by $500 million between July 1, 2017, and June 30, 2021.
5. Achieve 100% of energy use from renewable sources such as wind, solar, biogas, and geothermal by June 30, 2025 (City of Grand Rapids 2016).

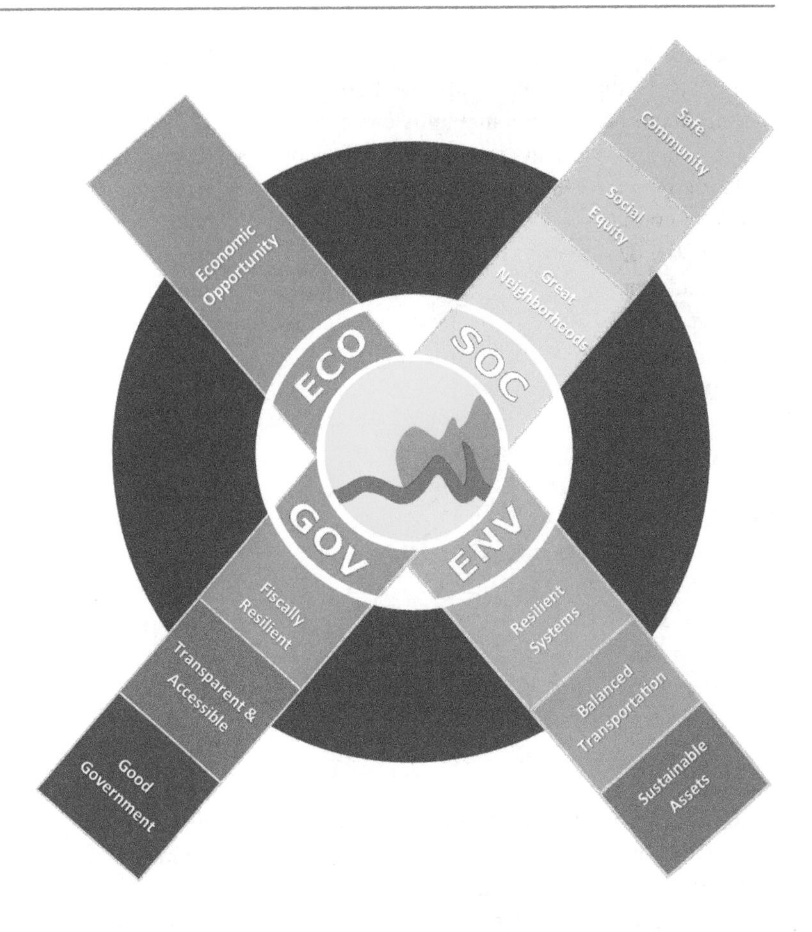

Fig. 4.2 City of Grand Rapids' quadruple bottom line pillars. (Source: City of Grand Rapids (2016))

A sample of the themes, goals, outcomes, and targets from the existing city's sustainability plan.

Theme 5: Resilient Systems (City of Grand Rapids 2016, p. 17).

Goal 1
Reduction of Energy Footprint

Outcome 1.1
Implement initiatives to counteract the effects of GHG emissions to provide a cleaner and greener community and a higher quality of life.

Targets

- Reduce the city's greenhouse gas (GHG) emissions to 25% below 2009 levels by 2021.
- Achieve 100% of energy use from renewable sources such as wind, solar, biogas, and geothermal by June 30, 2025.
- Double the water reuse and recovery by June 30, 2021 (from 360,000 gallons/day to 720,000 gallons/day).
- Ensure that – consistent with Grand Rapids' commitment to sustainability – the production, conversion, and renovation of committed affordable housing are on target with goals to reduce GHG emissions in the city.

Goal 2
Strengthen Climate Protection and Resiliency

Outcome 2.1
Integrate operations and preparedness measures into city plans to respond to climate change-related threats and disasters.

Targets

- Implement the climate resilience best practices in infrastructure planning by considering best available climate projections and incorporating climate adaptation planning into capital, operating, and maintenance programs by June 30, 2021.
- Expand opportunities for students to learn about, and take action on climate change by hiring four sustainability interns annually.
- Implement 100% onsite stormwater management to all new city infrastructure plans by June 30, 2025.

Goal 3

Expansion of Reuse and Recycling Opportunities

Outcome 3.1

Expand reuse and recycling opportunities, as well as composting of yard waste to decrease the amount of waste sent to landfills

Targets

- Reduce the number of landfill contributions by 10% within 5 years.
- Implement a recycling campaign for all city departments, and survey employees about recycling trends by June 30, 2021.

Theme 10: Good Government (City of Grand Rapids 2016, p. 24).

Goal 1

Provide Effective and Efficient Service Delivery

Outcome 1.1

Implement decisions, and follow processes that make the best use of resources to serve the needs of the entire community while balancing competing interests.

Targets

- Increase income tax receipts by an additional 5% over FY15 results by June 30, 2021.
- Increase the use of online permitting by an additional 25% over FY15 results by June 30, 2021.
- Respond to 100% of service requests for street maintenance from citizens, excluding potholes, capital improvement projects, and rehabilitation of infrastructure, within 72 h annually.
- Respond to 100% of street lighting outages within 72 h of being reported annually.
- Maintain 100% IT network security annually.
- Repair the emergency water main breaks, and restore service within 5 h at least 90% of the time by June 30, 2021.
- Ensure that 100% of sidewalk snow removal complaints will be abated within 72 h of the non-compliance notice expiration annually.
- Increase the city's overall fire code inspection completion rate to 90% by June 30, 2021.
- Increase the conversion of 311 phone/walk-in service to digital (self-serve) by an additional 25% by June 30, 2021.
- Decrease the number of walk-in customers by 20% by transitioning pay customers to the automated payment system.

Resilience Australian Cities Spotlight: Model of QBL Sustainability

The QBL method simplifies the objectives and provides for broad goals, objectives, and measurements. Similarly, sustainability and resilience plan progress reports follow the well-defined and identified structure of the planning process, contributing to a broader acceptance and appreciations of the plan. A number of local governments in Australia, including Liverpool, Stirling, Penrith, Norwood, and Lake Macquarie, have adopted a Quadruple Bottom Line in their sustainability policies or programs.

For instance, the city of Joondalup depicts sustainability as "meeting the challenge of striving simultaneously toward:

- Social responsibility – making decisions that lead to greater physical, cultural, and financial access and equity in service delivery and activities
- Environmental responsibility – not using more resources than required to deliver activities and services
- Economic responsibility – promoting and maintaining a city's economic development and growth in a sustainable manner
- Ethical responsibility – good governance, values, and behaviors (City of Joondalup n.d., 4)

Moreover, the city of Stirling is dedicated to "systematically review its internal policies, procedures, processes, and practices to further build the organization's capacity to deliver an ongoing quadruple bottom line (social, environmental, economic, governance) performance improvement" (City of Stirling 2009). In the city of Liverpool, the fourth bottom line "governance" of the integrated planning model is covered in the definition involving civic leadership and decision-making. A more in-depth reporting mechanism would further enhance the QBL practice to sustainability planning in these cities. Herriman et al. (2012) discovered the QBL plays an essential role in Australian cities' sustainability planning. A further cross-comparison on how the local governments in the United States and Australia address sustainability and resilience planning is warranted.

Objective Sustainability and Resilience Measurements

Sustainability and resilience actions are often viewed through a top-down lens, coupled with bottom-up participation for meeting the organizational objectives and goals. The most effective stakeholder engagement is accomplished with staff being part of the process of developing objective measurements and ultimately becoming champions for sustainability and resilience. With metrics and adequate staff engagement, employees have a better appreciation of their role in the sustainability and resilience planning initiatives undertaken by an organization. With proactive employees' engagement, measurements provide vital stakeholders with the critical information related to goals, achievements, and the fulfillment of an organization-wide vision. It is advantageous to associate the long-term vision and the organization's commitment toward sustainability and resilience. Having the visioning and planning components expressed separately creates a sense of disconnect and barriers to implementing sustainability and resilience objectives.

In monitoring sustainability and resilience, organizations inform the decision-makers and employees of the current operational capacity and performance and highlight areas in need of improvement, corrections, and adjustments. Conversely, demonstrating quantifiable successes toward sustainability presents an opportunity for positive encouragements in the workplace, leading to higher employee engagement, and ultimately job satisfaction. In implementing measurement systems with organizations, employees receive timely and essential feedback on progress and areas in need of further revisions and improvements. However, in using effective measurement techniques, organizations share success stories and adjust expectations, accept failures, and learn to adapt to changing environmental pressures.

Objective measurements include targets, a scorecard, a metrics, baseline year, benchmarks, and a sustainability report. Targets are the most impartial ways to measure sustainability and resilience initiatives. Targets represent a measure of the collective efforts. Furthermore, targets are indicators of organization's goals for proficient service delivery and organizational efficiency as compared to baseline year using benchmarks. While service delivery is the outcome of local government activities, the organizational efficiency is used for improving service delivery, reducing operational costs, and avoiding potential fiscal and budgetary pitfalls. For example, a local government may establish an energy efficiency target, coupled with a renewable energy target, displaying a reduction in energy consumption, and reinvesting savings into renewables. By reducing a total energy consumption, the local governments are able to meet their renewable energy target sooner and at a lower cost. Additionally, local government can create and display the reduced carbon footprint, as a direct result of a reduction in energy consumption and renewable energy production or purchase.

Showing the commitment to reducing energy consumption and production or purchase of renewable energy affords the city staff to adequately work toward reaching an exact goal and target expressed in the plan. Efficiency targets may be expressed in several ways. The energy conservation outcomes are articulated as an amount of energy reduced as compared to prior years, with a baseline year over a period of time, featured in kilowatt-hours (KWh) or percentages. Moreover, the particular energy targets are reported as savings in the amount of power saved or cost avoided. The targets for reducing energy consumption shown in currency/dollar amounts saved are adjusted for annual energy cost increases. Renewable energy target may be expressed in percentages of the total annual energy consumption for all the city operations, including facilities and processes. The renewable energy target may be offset with the existing renewable energy portfolio in the system, as states adopt the statewide renewable energy portfolio standards (RPS).

In some states with renewable energy portfolio standards, each power producer is required to produce a certain amount of megawatt hours (MWh) from renewables, allowing each power user to consume a portion of renewable energy in the portfolio of energy used. Alternatively, the renewable energy target could be expressed in KWh (kilowatt-hours), with a detailed amount of KWh the organization requires to achieve annually. Both renewable energy and energy efficiency targets are tracked using a baseline year and then with a targeted year in the future. A precise goal can be traced using a calendar or budget year. However, as most local governments have a mismatched calendar and fiscal year, ideal reporting mechanism for sustainability and resilience planning would align it with the fiscal plan calendar. Under ideal sustainability and resilience planning scenarios, each target would be tied to a budgetary process and be directly connected to the fiscal plan (Table 4.1).

Finally, empowering staff to take ownership of sustainability and resilience targets increases the level of collaboration and leadership among employees. The targets created organically, owned, and championed by employees and connected to the fiscal plan have a greater chance of attainment. Each champion of the sustainability and resilience target works closely with her or his counterparts in other departments to ensure the progress and ultimate realization of the target. The facilitation and the involvement of city departments at different organizational levels foster the embedment of sustainability and resilience within an organization.

Practical Application: Starting Simple and Small

A vital element of sustainability and resilience planning is aligned with the local government's energy efficiency and conservation initiatives. The history of energy conservation in cities goes back to first facility audits which were conducted, implementing basic cost reduction strategies and evolving into the modern energy efficiency strategies. Energy conservation efforts remain a centerpiece of the overall sustainability and resilience initiatives of many local governments, as they are

Table 4.1 Creating a specific target for sustainability and resilience plan

2020 renewable energy target	2025 renewable energy target	2030 renewable energy target	2020 energy efficiency target	2025 energy efficiency target	2030 energy efficiency target
50% of the total energy use	75% of the total energy use	100%	Reduce consumption by 1% over the 2019 baseline year	Reduce energy use by 2% over 2019 baseline year	Reduce 3% over the 2025 energy consumption

manageable, easy to measure, track, substantiate, and demonstrate an immediate return on investment. Moreover, numerous local governments have a renewable energy target, whether to produce or to procure green energy. Staff dedicate employees to seek innovative solutions for reducing the energy consumption and reinvesting the energy savings into renewable energy projects.

To illustrate the commitment to sustainability and to increase resilience through reduced energy consumption, one of the initial steps within organizations is to track, measure, and report results of sustainable energy activities. With the establishment of the inventory of electricity use for all the city-owned buildings, local governments are able to comprehend the effects of energy use on the operations. By measuring, tracking, and reporting data, administrators can point out at potential opportunities for cost-saving measures. As an example of reporting on energy efficiency efforts, local governments demonstrate significant energy reduction, expressed as either direct savings or cost avoidance. Investments made in energy efficiency improvements provide an obvious payback. Furthermore, the local government administrators can highlight the CO2 equivalent removed in operations as a result of energy efficiency reduction and investments in renewables.

A sample of energy consumption reduction, from the baseline year of 2017 to the year of 2018, benchmarked against the previous year's data, is featured in Table 4.2, with a decrease of 1000,000 KWh. The effects of sustainability initiatives are immediately apparent. The indirect or direct savings are expressed in dollars and carbon reduction. The cost avoidance is a preferred term. It allows cities to put funds aside and reinvest them in energy efficiency improvements or renewable energy projects. Cost avoidance signals the expenses are ongoing and are continually accounted for in the present and in the future of operations.

Using a GHG emission calculator available at the Environmental Protection Agency (EPA) website, administrators may show their contribution to mitigating the impact of climate change by reducing carbon footprint and calculating the CO_2 equivalent saved. In the Table 4.2 example, 744 metric tons of carbon avoided from being released into the atmosphere from energy consumption is equal to 159 passenger vehicles driven for a year or 80.4 of homes powered for a year (EPA 2017). By measuring, tracking, monitoring, and then reporting out the progress, local governments garner support from constituents for the sustainability and resilience activities. Finally, while Table 4.3 provides examples of energy project, with cost avoidance, energy savings, and environmental carbon footprint reductions. Table 4.4 features a sample of renewable energy benefits.

Table 4.2 A sample of energy efficiency savings

2017 (KWh)	2018 (KWh)	Annual energy saved (KWh)	Cost avoidance at avg. $0.10/KWh	CO_2 equivalents saved (metric tons)
110,000,000	109,000,000	1000,000	100,000	744

Table 4.3 Sample of energy project, with cost avoidance, energy savings, and environmental carbon footprint reductions

Type of project	Cost	Annual utility savings (kWh)	Total cost avoidanc/ savings	GHG reduced (metric tons)
Window and door replacement	$1,200,000	251,911	$39,975	195
Lighting replacement	$330,000	407,337	$34,623	293
Occupancy sensors	$37,750	86,822	$7379	62
Geothermal projects	$300,000	266,000	$22,610	191

Table 4.4 Reporting renewable energy purchase in annual reports

% of renewable energy purchased (of total power)	kWh/year	Biogas	Wind	Solar	CO2 equivalents saved through green energy purchasing (metric tons)
25.75%	25,948,200	92%	7.5%	0.5%	17,893

Resilient City Spotlight: Ann Arbor – Sustainability Framework and Sustainability Action Plan

While not strictly a sustainability or resilience plan, the city of Ann Arbor's sustainability framework is a part of the city's master plan which showcases sixteen sustainability goals (City of Ann Arbor 2013). Ann Arbor's sustainability framework integrates over 200 sustainability-related action items, policies, and programs from 27 plans into a single overarching set of 16 sustainability goals divided into four categories: climate and energy, community, land use and access, and resource management (City of Ann Arbor 2013). Likewise, environmental goals have one or more indicators used to measure progress. A stable climate is one of the environmental goals with greenhouse gas (GHG) emissions; vehicle miles traveled (VMT), electricity use, and natural gas use are listed as indicators (City of Ann Arbor 2013). Counterintuitively, the city's sustainability report is entitled sustainability action plan with 38 indicators to showcase progress (City of Ann Arbor 2015) (Fig. 4.3).

The city's interactive website offers a comprehensive overview of the sustainability progress supported by engaging interactive tools. Those interested in more details can view each indicator with specific statistics, data, and supplementary information for the related sustainability initiatives. Using a color-coded system for sustainability dashboard, administrators designate the color for each indicator: green for good, yellow for fair, red for poor, and white for not assessed. Furthermore, each indicator with an upward pointing arrow represents an indicator getting better, a horizontal arrow for stable, a downward pointing arrow for getting worse, and a question mark for unknown (City of Ann Arbor n.d.). It is a visually attractive system and an entertaining method to engage the viewers of the results (Fig. 4.4).

Examples of the quantifiable goals in the report are featured under Resource Management Responsible Resource Use. One of the goals is to "produce zero waste and optimize the use and reuse of resources in the community," with corresponding actions for each area of the city's planning to meet the goal (City of Ann Arbor 2015, p. 28).

Target RM 7: Increase waste diversion rates from 50% to 60% for single family residence by 2017 (Solid Waste Resource Plan). Activities include:

- Explore the option of increasing city compost collection from seasonal (April to mid-December) to year-round (Solid Waste Resource Plan).
- Distribute 5000 kitchen composters by June 30, 2014 (Systems Planning Budget Goal).
- Provide expanded food waste composting service to all curbside collection routes (Solid Waste Resource Plan).
- Increase the sale of compost carts by 10% (Field Ops Budget Goal).

Target RM 8: Increase citywide waste diversion rates up from current 31% to 40% by 2017 (Solid Waste Resource Plan). Actions include:

- Increase recycling participation through pilots, such as a recycling incentive program for multifamily units (Solid Waste Resource Plan).
- Place 15 new recycled dumpsters (Field Ops Budget Goal).
- Expand types of materials collected in the city recycling program as markets and processing abilities develop (Solid Waste Resource Plan).
- Expand away-from-home recycling opportunities (Solid Waste Resource Plan) (City of Ann Arbor 2015, p. 28).

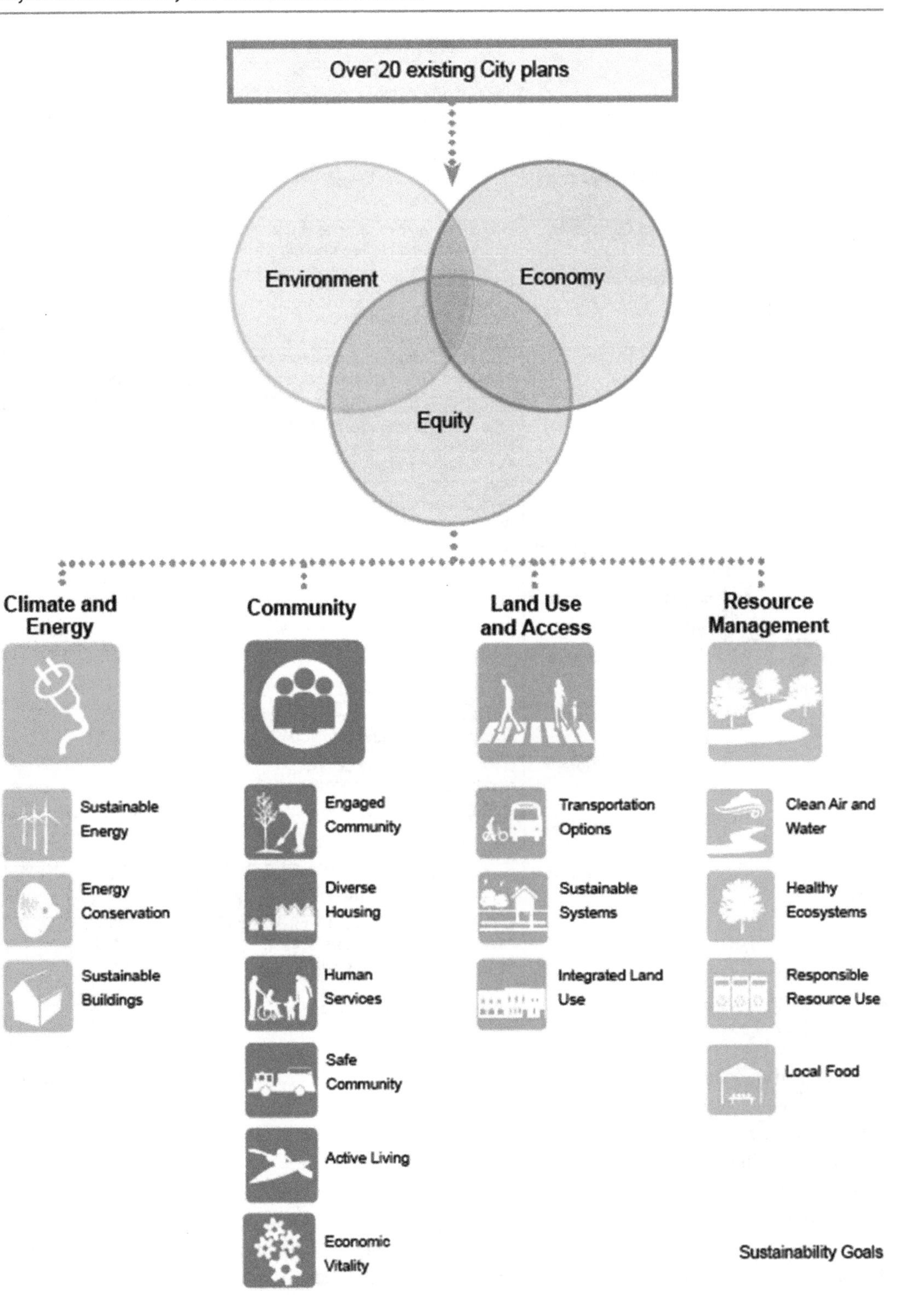

Fig. 4.3 City of Ann Arbor (2013) sustainability framework goals

	Goal	Indicator(s)
Climate and Energy	Sustainable Energy	Renewable Energy Generation Number of Renewable Energy Projects
	Energy Conservation	Greenhouse Gas Emissions
	Sustainable Buildings	Dollars Invested in Energy Efficiency Loans LEED Buildings
Community	Engaged Community	Opportunities to Participate in Community Matters (NCS) Attended a Local Public Meeting (NCS)
	Diverse Housing	Percent of Public Housing Meeting Enterprise Green Communities Standards Percent Affordable Housing Availability of Affordable Quality Housing (NCS)
	Human Services	Percent of Public Housing Residents Employed Percent of Public Housing Units with Supportive Services
	Safe Community	Structures in the Floodplain Emergency Preparedness (NCS) Police Services (NCS) Fire Services (NCS) Part 1 Crimes
	Active Living and Learning	Recreational Opportunities (NCS) Social Events and Activities (NCS)
	Economic Vitality	Overall Economic Health (NCS) Value of Construction Permits
Land Use and Access	Transportation Options	Overall Ease of Travel (NCS) Percent Non-Motorized Commutes
	Sustainable Systems	Tree Condition Impervious Surface Average Pavement Rating
	Integrated Land Use	Land Use, Planning, and Zoning (NCS) Overall Built Environment (NCS)
Resource Management	Clean Air and Water	Days of Unhealthy Air Quality Number of EV Charging Stations Green Infrastructure Capacity
	Healthy Ecosystems	Tree Canopy Watershed Health
	Responsible Resource Use	Percent Waste Diverted Recycling Collection (NCS)
	Local Food	SNAP/EBT Purchases at Farmers Market Greenbelt Land Preserved Availability of Affordable Quality Food

Fig. 4.4 City of Ann Arbor (2015) a summary of action plan indicators, p.3

Goals and indicators are identified, and measurements are in place. Short of directly determining the target champions, the city staff associated the indicators and activities with the documents in the city's operational strategies, such as plans or budget goals are assigned the duty to meet the specified goal and objectives of the plan. An example of nonquantifiable descriptive goal in the report is in the same section of resource management, under air quality (City of Ann Arbor 2015, p. 24).

Target RM 2: Promote green transportation improvements to reduce vehicle emissions (Transportation Plan). Actions include:

- Establish requirements for electric vehicle (EV) parking infrastructure for projects, and increase citywide infrastructure for EV charging (Climate Action Plan).
- Add more EV car charging stations and infrastructure in the downtown (Downtown Development Authority). Support idling edu-

cation program through signage and other materials (Environmental Commission).

- Develop evaluation techniques to gauge effectiveness of previous and current Ann Arbor Safe Streets and Sidewalks (A2S3) Committee education campaign strategies, and recommend new outreach ideas (non-motorized transportation plan, planning and policy updates).
- Use the city's geographic information system (GIS) to provide the current representation of the biking and walking facilities in the city (non-motorized transportation plan, planning and policy updates).

While more descriptive, these goals and corresponding indicators serve as strategies to achieve particular outcomes.

Reporting

Local governments implementing sustainability and resilience report the final and interim results and incorporate them into the organizational goals and objectives. By presenting outcomes from sustainability and resilience implementation, a local government organization exercises a democratic right to communicate to constituents what it stands for, what needs to be improved, and the future plans. Staff collect data and then convert the data into tables, graphs, and charts to communicate results. Benchmarking results with prior years allows for goal setting for future years.

Practical Applications: Cities of Fort Collins and Grand Rapids Progress Reports

An example of a simple yet powerful way to communicate progress is designed by the local government of the city of Fort Collins, CO, featured in their annual sustainability report. The city features the relevant measures, showing the difference between the baseline year of 2005 and benchmark year of 2013 for the key city indicators. The local government of Fort Collins demonstrated meaningful reductions and progress in what the local administrators described as "a per capita or square foot measurement, based on emission sources (City of Fort Collins 2013):

"Total CO2e emissions: 7.6% decrease;

- Per 1000 sq. ft. CO2e emissions: 4 metric tons;
- Per employee: 5 metric tons, a 37% decrease;
- Per vehicle emissions: 0.8 metric tons decrease;
- Electricity generated by clean, renewable energy onsite: 36 kW increase;
- Change in tons of waste sent to the landfill: 52% decrease;
- Carbon emissions from electricity: 9.4% decrease;
- Conventional fuel use: 19% decrease;
- Electricity used for traffic signals: 42% decrease;
- Electricity use for water and wastewater production: 5.6% decrease" (City of Fort Collins 2013, p. 2)

Furthermore, an in-depth report demonstrates the Scope 1 (direct carbon emissions, natural gas fuels), 2375 metric tons CO2e increase; Scope 2 (energy indirect emissions, electricity), 3490 metric tons CO2e decrease; and Scope 3 CO2e emissions (gases from waste to landfills, recyclables, personal vehicle travel, and air travels), 3510 metric tons CO2e decrease (City of Fort Collins 2013 p. 2) (Fig. 4.5).

Furthermore, local governments may measure, track, monitor, and then report each energy efficiency target or each project separately and report it in the annual report as overall progress toward meeting a specific energy target. The local governments feature total progress in all areas of sustainability and resilience, by showing how many specific targets or indicators are met, and and how many are making progress or not making progress at all.

Overall results in the pie chart in Fig. 4.6 indicate the total percentage of the targets met as part of the total number of targets adopted by the local government. In this case, the local government

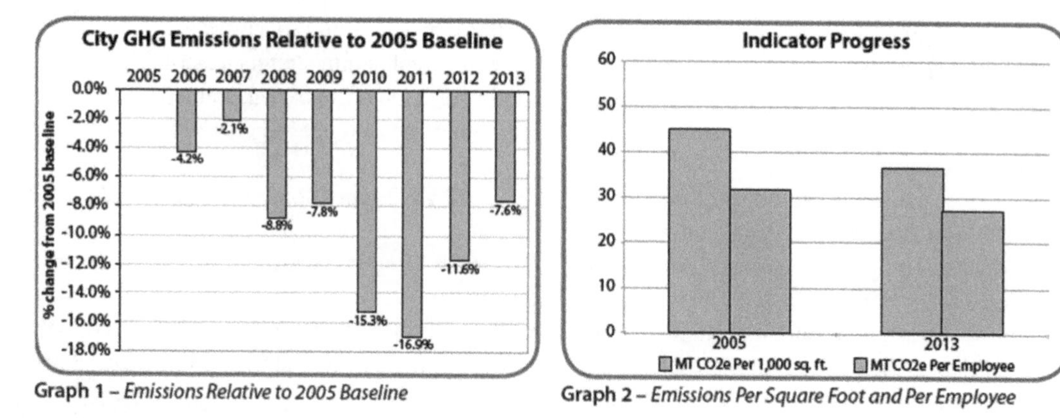

Fig. 4.5 City of Fort Collins, 2013, p. 2, indicator progress

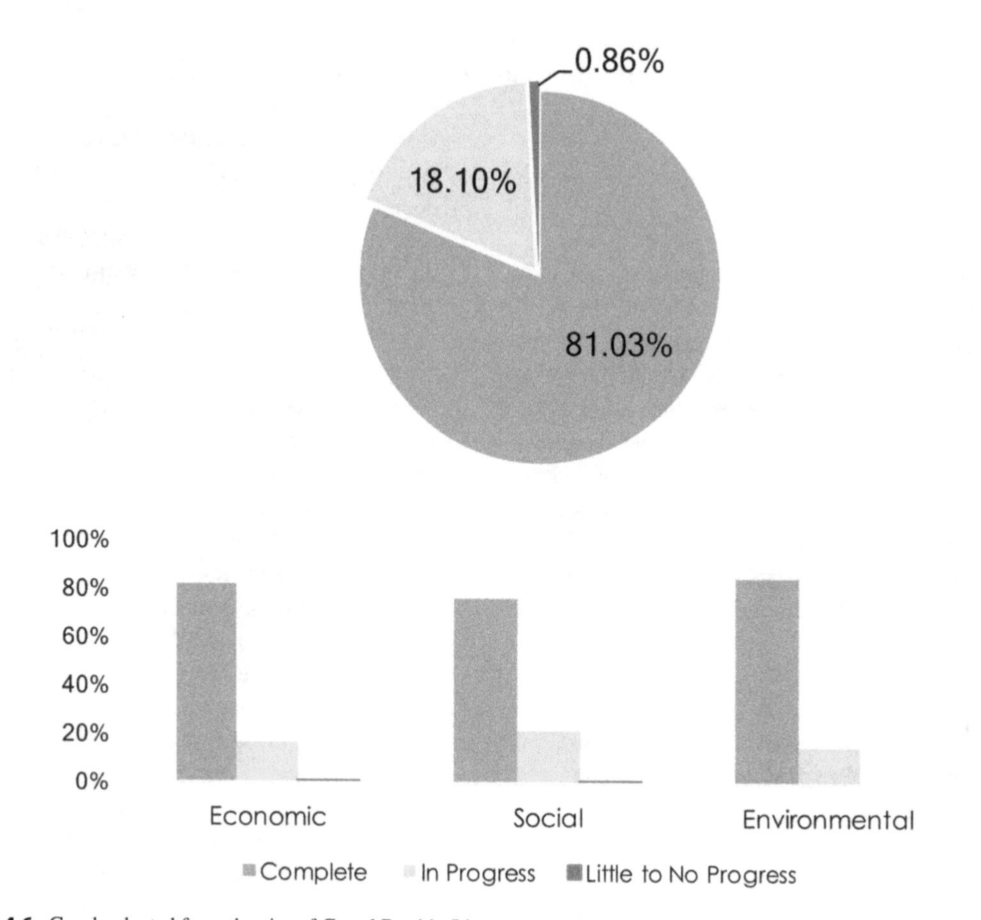

Fig. 4.6 Graph adapted from the city of Grand Rapids 5th year annual report

has met 99.1% of the sustainability targets or made progress after the final year of plan implementation. Of the total number of targets, local government has completed an 81.0% of all the targets as set in the sustainability plan. Furthermore, these targets can be compared to a previous year's results and can be broken down to a theme, goal, or an objective (Fig. 4.7).

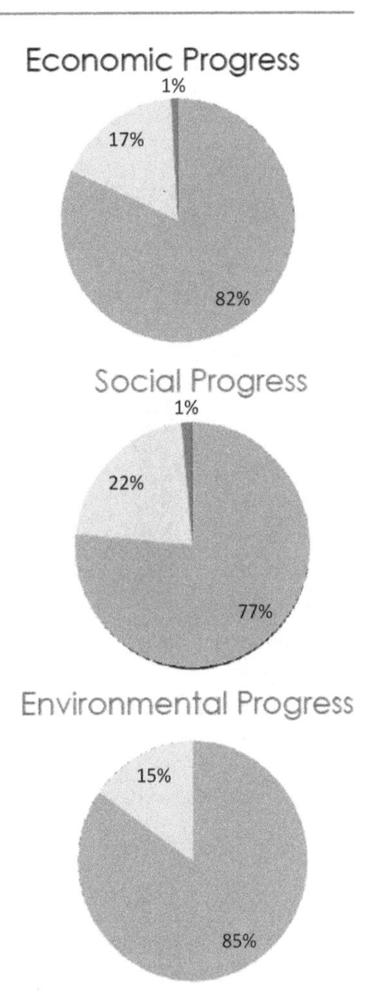

Fig. 4.7 Report adapted from Grand Rapids annual sustainability report

Administrators use reports to feature success stories and share developing practices, implementation of sustainability and resilience initiatives, cost avoidance, savings, and to provide information about future steps and projects (Figs. 4.8 and 4.9).

Local government can break down the report into more details, showing the sources of renewable energy. Additionally, in their annual report, the city of Fort Collins staff, in visually appealing presentation, call to attention game-changing projects, projecting both the environmental benefits and estimated savings (City of Fort Collins 2013).

Reporting is a crucial piece of the sustainability and resilience puzzle. With correct reporting, local governments feature results and are transparent and accountable to the local constituency. A notable component of the cities' engagement in sustainability and resilience is attributed to the sustainability plan and annual reporting. The local government plans serve as a multiyear, adaptable blueprint used by each department to plan activities and justify budget requests based on economic, social, environmental, and governance related outcomes.

Centralization of Reporting

A review of the various sustainability and resilience plans, policies, and programs in local governments revealed the significance of having a single location or a point of contact for measuring, monitoring, tracking, and reporting results

Energy- Reduction

Since FY2009, the City has reduced electricity usage by 5.53%, a reduction of over 5,800,000 kWh.

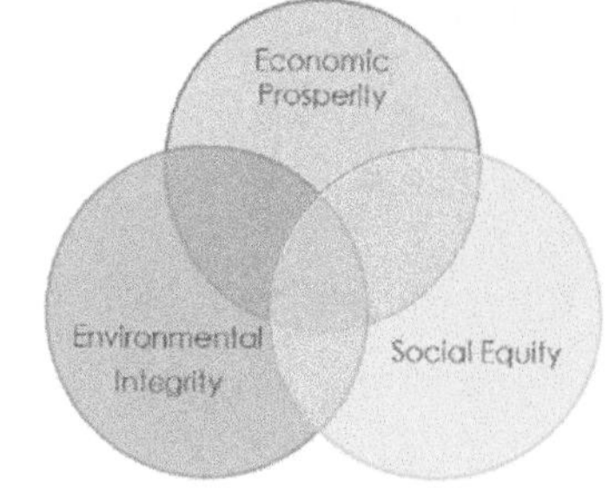

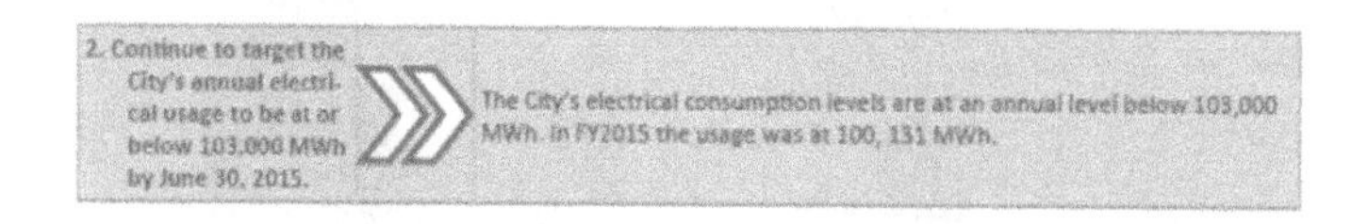

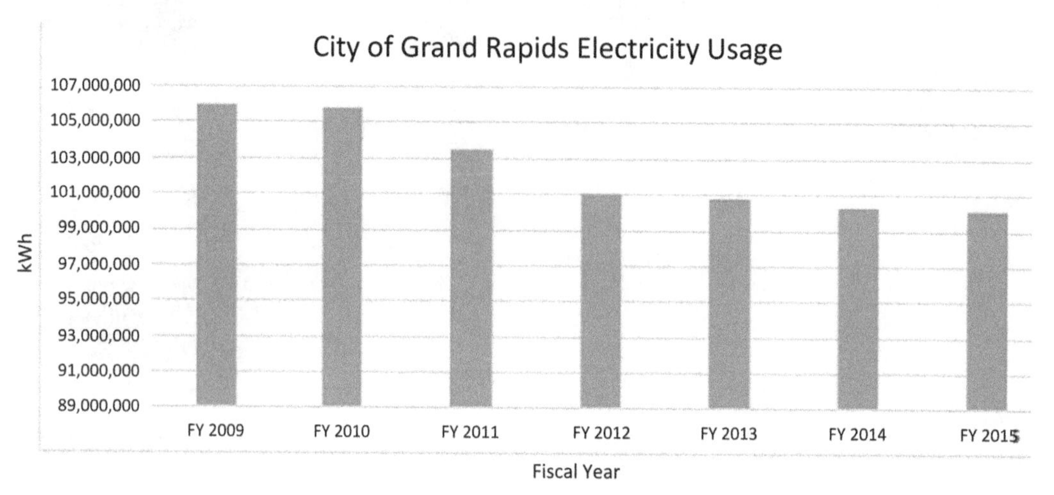

Fig. 4.8 Graph adapted from the city of Grand Rapids' 5th year progress report

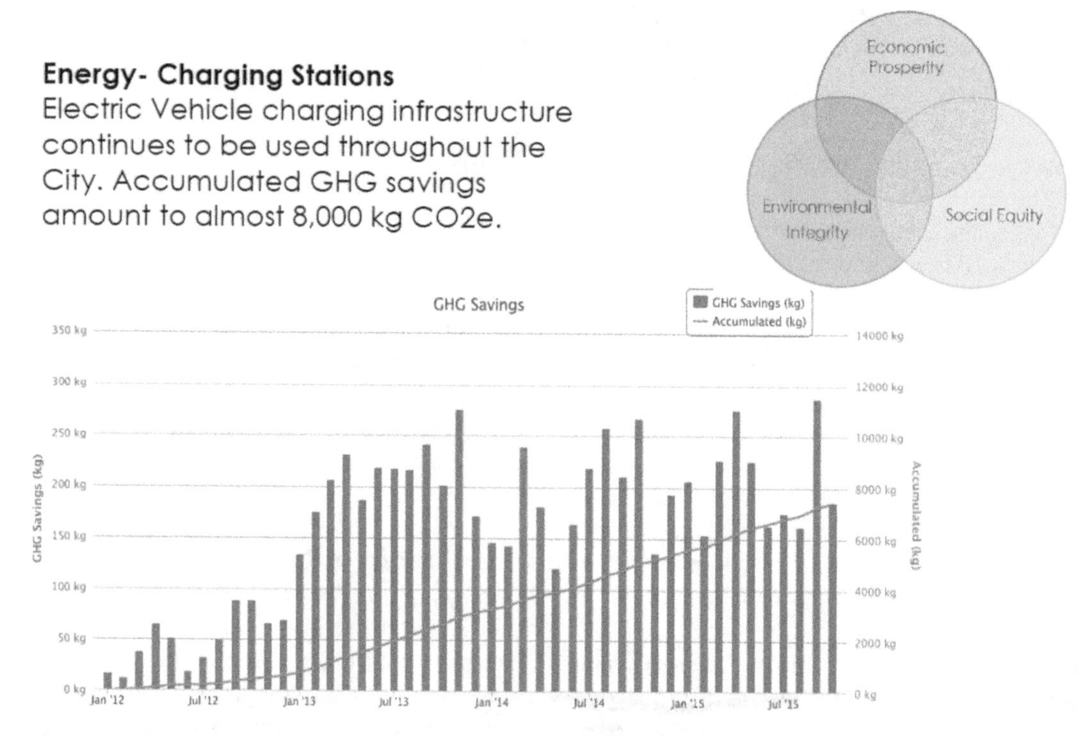

Fig. 4.9 Graph – reporting on specific projects. (Adapted from the city of Grand Rapids, sustainability plan progress report)

and outcomes. Whether it's the Office of Energy and Sustainability in Grand Rapids, Ann Arbor Sustainability Office, Office of Sustainability in Denver, or the Sustainability Services in Fort Collins, CO, a central and unified front exists to compile the reports, making the efforts more focused. While cities and their sustainability and resilience initiatives are different in nature and outcomes, they point out to the relevance of consistency and ownership of the process of reporting the sustainability and resilience activities. Having a central point of contact for data collection and reporting of all the sustainability and resilience activities enables other staff to continue contributing to sustainability and resilience and, at the same time, focus on completing their daily tasks. Sustainability and resilience planning can become an integral part of the budgeting process, through active pursuit of sustainability goals and targets.

Cities have an enormous impact on the economy, society, and environment through decisive actions aimed at reduction of the energy demand and energy consumption. Local governments are faced daily with crucial decisions on providing services and meeting increased demand for services while facing constant and severe budget cuts to staffing and operations. At the same time, city governments are expected to provide the same level of services without additional revenues or resources. Institutionalizing sustainability and resilience is an enormous undertaking, which requires leadership and readiness to measure, track, and report progress. Tracking, measuring, monitoring, and reporting are critical to success for all local government projects. Data from each department can be collected and disseminated through a single point of contact, utilizing a single point person and sustainability and resilience office. With reporting, the department is still responsible for collecting data. However, the reports are done on a macro-level to reveal the city's energy consumption more noticeably. When available funding is in peril, local governments' capacity to pursue sustainability and resilience becomes an added effort, beyond required tasks to provide essential services. Sustainability and resilience planning becomes an integral part of the budgeting process, through active pursuit of sustainability and resilience goals and targets using the Quadruple Bottom Line approach to reporting (Table 4.5).

Table 4.5 Benefits of centralized reporting – framework for sustainability and resilience progress updates

QBL Benefits	centralized reporting
Carbon reduction/ environmental	Health benefits, reduced pollution, reduced GHG emissions/CO2 emission reductions, cleaner rivers and water, regional air quality benefits, climate change, fossil fuel reductions
Systems/ societal	Sustainable operations, clean energy, energy change, resilience, resilient communities and organizations
Governance/ operational	Cost cutting, cost avoidance, price reduction, KWh saved, savings, frugality, answerability, transparency, good governance, stewardship, return on investment, accountability, responsibility, responsiveness, budget alignment, fiscal resiliency
Economic	Job creation, transformation, city attractiveness, economic development, business support,

Summary

Local governments engaged in sustainability and resilience planning adapt, transform, and accept discontinuities as they continue to deliver services without interruptions. In a crisis, sustainability and resilience-related initiatives convert into an opportunity and a tool for local governments to change priorities and outcomes of the budget process. The approach of a comprehensive accountability and transparency provides the primary direction for the goals, outcomes, and targets established in a sustainability and resilience plan. The Quadruple Bottom Line approach to measuring sustainability and resilience outcomes ensures the local government staff are being held accountable. Progress reports available at the local governments' websites provide additional transparency and an opportunity for community engagement. Local governments measure and report targets regularly. Measuring activities allows administrators to value the outcomes and goals being measured and tracked. Measurement allows for complete accountability and transparency of the implementation of sustainability and resilience initiatives. The design of the sustainability and resilience programs does not necessarily lead to more efficiency unless it is intended to be tracked, measured, and reported.

The foundation for sustainability and resilience vision is the ability to measure, monitor, track, and report the sustainability and resilience initiatives, programs, and outcomes. Once a sustainability and resilience target achievement is reported, measured, and then compared to various outcomes, it assists organizations to assess the positive impact on the overall effectiveness of service delivery of local government. Progress reports availability at local governments' websites is crucial for transparency of sustainability and resilience-related efforts. Targets showing progress in areas of sustainability and resilience should at minimum be reported annually and be tied to an annual fiscal plan and budgeting processes.

Local governments can operationalize the sustainability and resilience metrics. By directly tying sustainability and resilience initiatives to the annual budget, the sustainability and resilience activities are centralized and embedded. The centralization of reporting through a single department is critical. Reporting furthers the transparency, accountability, answerability, and good governance of the local government operations. Local governments use positive, practical examples of sustainability and resilience-related projects to improve the overall governance of their operations. Sustainable energy initiatives function as the essential platform for organizational engagement in sustainability and resilience efforts. In encouraging the ownership of sustainability and resilience targets, city staff are held accountable and are in control of sustainability and resilience initiatives.

Further Discussions

- Assess the importance of measuring, tracking, and reporting the sustainability and resilience targets.
- Explain the points of differences between indicators and targets.
- Discuss various approaches local governments undertake in defining and reporting measurable outcomes and a sample of targets and corresponding measurements for sustainability and resilience efforts.
- Review local governments' annual sustainability and resilience reports.
- Analyze and compare the metrics, goals, targets, baseline, sources of data, and other areas of reporting.

References

Alachua County (2008) Declaration for resource efficient and resilient county. Retrieved from http://www.alachuacounty.us/Depts/Manager/Sustainability/Documents/declaration.pdf

Alibašić H (2017) Measuring the sustainability impact in local governments using the quadruple bottom line. Int J Sustain Policy Pract 13(3):37–45

Alibašić H (2018) Ethics and sustainability in local government. In: Farazmand A (ed) Global encyclopedia of public administration, public policy, and governance. Springer International Publishing AG, Cham. https://doi.org/10.1007/978-3-319-31816-5_3427-1

Bell S, Morse S (2008) Sustainability indicators: measuring the immeasurable? Earthscan Publishing, Washington, DC

Brugmann J (2012) Financing the resilient city. Environ Urban 24:215–232. https://doi.org/10.1177/0956247812437130

City of Ann Arbor (2013) Sustainability framework – SustainA2ble: cultivating our people, place, and potential. Retrieved from. https://www.a2gov.org/departments/systems-planning/planning-areas/climate-sustainability/sustainability/Documents/Ann%20Arbor%20Sustainability%20Framework%20051313.pdf

City of Ann Arbor (2015) Sustainability action plan. Draft July 2015. Retrieved from https://www.a2gov.org/departments/systems-planning/planning-areas/climate-sustainability/Sustainability-Action-Plan/Documents/SAP%20-%20DRAFT%20-%20%20July%202015%20-%20web.pdf

City of Ann Arbor (n.d.) Sustainability dashboard. Retrieved from: https://www.a2gov.org/departments/systems-planning/planning-areas/climate-sustainability/Sustainability-Action-Plan/Pages/Dashboard.aspx

City of Cincinnati (2013) Green Cincinnati plan. Retrieved from https://www.cincinnati-oh.gov/oes/linkservid/6CE53223-9206-9F36-DB7FA3444F16A1A0/showMeta/0/

City of Fort Collins (2013) 2013 Municipal sustainability annual report. Retrieved from https://www.fcgov.com/sustainability/annualreports/2013-report.pdf

City of Grand Rapids (2015) Fifth year sustainability plan progress report. Office of energy and sustainability. [Alibašić H and Gosztyla D (eds)]. http:// www.grandrapidsmi.gov/files/assets/public/departments/office-of-sustainability/reports-and-documents/office-of-sustainability/5th-year-progress-report-sustainability-plan.pdf.

City of Grand Rapids (2016) Sustainability plan FY2017-FY2021. Office of energy and sustainability. [Alibašić H (eds)]. http://www.grandrapidsmi.gov/files/assets/public/departments/office-of-sustainability/reports-and-documents/office-of-sustainability/sustainability-plan-fy17-fy21.pdf

City of Joondalup (n.d.) Policy 5–4 sustainability. Retrieved from http://www.joondalup.wa.gov.au/Libraries/Policies/Sustainability_5-4.pdf

City of Stirling (2009) Sustainability policy. Retrieved from: https://www.stirling.wa.gov.au/Council/Policies-and-local-laws/Policy%20and%20Local%20Laws/Sustainability%20Policy.pdf

Doppelt B (2010) Leading change toward sustainability. Greenleaf Publishing Limited, Sheffield

Elkington J (2012) The Zeronauts: breaking the sustainability barrier. Routledge, New York

Hemmati M, Enayati J (2002) Multi-stakeholder processes for governance and sustainability: beyond deadlock and conflict. Earthscan, London

Herriman J, Hazel S, Phil S, Grahame C (2012) Working relationships for sustainability: improving work-based relationships in local government to bring about sustainability goals. Commonw J Local Governance. 10:116–33. Retrieved from http://epress.lib.uts.edu.au/journals/index.php/cjlg/article/view/2693/2906

Hitchcock DE, Willard ML (2008) The step-by-step guide to sustainability planning: how to create and implement sustainability plans in any business or organization. Earthscan, London

Kennedy S, Sgouridis S (2011) Rigorous classification and carbon accounting principles for low and zero carbon cities. Energy Policy 39(9):5259–5268. https://doi.org/10.1016/j.enpol.2011.05.038

Liverpool City Council (2012) Integrated environmental sustainability action plan. Retrieved from http://www.liverpool.nsw.gov.au/trim/documents?RecordNumber=089986.2016

Liverpool City Council (n.d.) Our vision for sustainability. Retrieved from http://www.liverpool.nsw.gov.au/environment/sustainability

Mascarenhas A, Nunes LM, Ramos TB (2015) Selection of sustainability indicators for planning: combining stakeholders' participation and data reduction techniques. J Clean Prod 92:295–307

Pacioli FL (1494). Particularis de computis et scripturis (Accounting books and records from from Pacioli's summa de Arithmetica Geometria Proportioni et Proportionalita) (J. Cripps trans.). Pacioli Society, Seattle

United States Environmental Protection Agency (EPA) (2017) Energy and environment. Greenhouse gas equivalencies calculator. Retrieved from https://www.epa.gov/energy/greenhouse-gasequivalencies-calculator

West Michigan Environmental Action Council (WMEAC) (2013). Grand rapids climate resiliency report. Retrieved on September 11, 2017, from: https://wmeac.org/wp-content/uploads/2014/10/grand-rapids-climate-resiliency-report-master-web.pdf

Yukl G, George JM, Jones GR (2010) Leadership: building sustainable organizations (laureate education, Inc., custom ed.). Custom Publishing, New York

Implementing a Sustainability and Resilience Plan: Initiatives and Programs

5

"Cities require a concentration of food, water, energy and materials that nature cannot provide. Concentrating these masses of materials and then dispersing them in the form of garbage sewage, and as pollutants in air and water is challenging city managers everywhere." Page 206, Brown, Lester R. (2006). Plan B 2.0: Rescuing a Planet Under Stress and a Civilization in Trouble (Updated and Expanded). New York, NY: Earth Policy Institute

Key Questions

The fifth chapter of this book aims to answer the following underlying assumptions and queries:

- What are sustainability and resilience-related activities, initiatives, and outcomes?
- What type of sustainability and resilience programs and projects work?
- How do cities select particular programs and projects?
- How does a preemptive investment in sustainability and resilience planning provide a return on investment?
- Who champions and implements the plans?

Introduction

The embedment of sustainability and resilience at all levels of local government is an arduous, holistic, and dynamic process. Thus far we covered several vital components contributing to the achievement of sustainability and resilience outcomes, including but not limited to:

- Internal and external scan, stakeholder identification, and operational capacity
- Meaningful stages of policies and program toward creating and implementing a sustainability and resilience plan.
- Internal and external stakeholder engagement and pursuit of partnership opportunities in seeking sustainability and resilience outcomes.
- Empowering employees to define and champion sustainability and resilience targets.
- Measuring, tracking, monitoring, and reporting sustainability and resilience targets and results and creating and sharing a sustainability and resilience plan progress reports.

Organizations design complex projects, components, and activities to implement sustainability and resilience. Local governments are committed financially and otherwise to the sustainability and resilience goals and objectives and the practical application and implementation of sustainability and resilience policies. By connecting sustainability planning with resilience and directly to the budget process, organizations indicate and substantiate commitment to those efforts.

From Plans to Applied Sustainability and Resilience

In recent years, local governments showcased the resolve to both plans and investments in infrastructure, to design the framework and imple-

H. Alibašić, *Sustainability and Resilience Planning for Local Governments*, Sustainable Development Goals Series, https://doi.org/10.1007/978-3-319-72568-0_5

ment strategies for sustainability and resilience. Local governments continuously face the risks of the economic recession, elementary exposures, extreme weather, and climate change. Cities, counties, townships, and villages invest preemptively into sustainability and resilience projects and programs to counter internal and external threats and pursue opportunities. The imperative in implementing sustainability and resilience in organizations is an inappropriate balance between the organizational objectives, vision, and effectiveness in delivering services.

Municipalities may evaluate the implementation mechanism of sustainability and resilience planning through an active assessment of the investments in infrastructure improvements. How well do the streets and road infrastructure perform? How many miles of sidewalks are missing along significant streets? How many miles of new bike lanes did the local government add in recent years? How many people were engaged in various community events and activities? What are the economic, environmental, social, and governance benefits from investments in sustainability and resilience? What are the actions steps to implementing low-impact design and engaging businesses to adopt emerging practices to reduce stormwater runoff? What benefits does an organization achieve from investments in sustainable energy? Besides, local governments have benefited from sustainable energy planning. Local governments link the implementation of sustainable energy initiatives to a fiscal plan to maximize gains for the organizations. Implementing sustainability and resilience is the most effective long-term strategy when environmental, social, economic, and governance advances are prominently featured, in a systematic and purposefully defined approach. Commonly, the local governments' approach to implementation of sustainability and resilience includes the following features.

Comprehensiveness, System, and Project-Driven Processes Vetting each project and integrating it into the system-wide plans. In a reporting scheme, complete with target champions, each target in a resilience and sustainability plan links to a single project. As an illustration, in strategies to implement energy objectives, power use or power production, it is each project that contributes to the overall energy reduction, to meet renewable energy target, and carbon footprint reduction.

Targets Are Quantified and Budgetary and Policy Aligned Each target is designed with a purpose and directly aligned to the annual fiscal and budgetary process and also to eventually provide for revisions and adjustments to existing plans and policies.

Answerability and Transparency Local governments' transparency and accountability synergies are directly related to their aptitude to implement and execute flexible, durable, sustainable, and resilient strategies, with regular reports, updates, and robust community engagement.

The Organizational Dynamic of Project Implementation

Each organization has a specific set of goals and objectives related to sustainability and resilience. Hardjono, Van Marewijk, and de Klein created the European Corporate Sustainability Framework (ECSF), who (as cited by Stubbs and Cocklin 2008) developed "a set of models, tools, and theories—to help organizations address complex social and environmental sustainability issues" (p. 104). Moreover, Hardjono et al. (2004) provided a comprehensive methodology for implementation of sustainability in organizations by capitalizing on organizational dynamics. The organization can deploy symbolic interaction methodology in a systematic approach to employ sustainability. The application of sustainability and resilience is accomplished by addressing all interconnected aspects of interests for the organizations, including but not limited to governance, economic growth and development, social impacts, and environmental concerns.

The organizational dynamics is relevant to all features of implementation strategies. As organizations attempt to address the economic, environmental, and social issues of projects, the often neglected portion of project implementation is the good governance. Alibašić (2017a) argued for local governments and other organizations to use the Quadruple Bottom Line (QBL) in the implementation of sustainability and resilience strategies to improve service delivery, transparency, and accountability among many other benefits from adding the fourth pillar to planning (p. 41). Additionally, good governance is one of the most critical elements of sustainable and resilient cities in the design and management of local governments' operation with the lowest carbon footprint and impact.

In their operational planning and long-term strategies, the local government administrators share common concerns and apprehension about the resource management and service delivery. Local government administration plays a precarious role in managing resources appropriately for a resilient and sustainable community. Municipal leaders and administrators utilizing sustainability and resilience planning at a minimum include the following elements in plans: robust and flexible economic, energy, water, natural environmental, transportation, and waste systems. Such systems are not resource dependent and are renewable resource based; and are efficient, resistant, and durable. The primary objective of organizations with sustainability and resilience planning in place is the improved conservation and use of resources. Furthermore, local governments seek reduced costs of operations, efficiency measures, and meeting of the economic, social, environmental, and good governance goals for the organization.

Quadruple Bottom Line (QBL) Implementation Objectives

The four distinct areas of implementation strategies are featured as social, economic, environmental, and governance pillars of resilient and sustainable community planning.

Prosperous and Resilient Economy

- Advancing drivers of economic development and growth.
- Stimulating the local and regional economic growth opportunities by supporting the job creation, updating and redeveloping neglected properties, removing the barriers to business development, and favoring environmentally responsible and socially conscious industries.

Social Equity and Resilient Communities

- Addressing the equitability, fairness, social justice, racial divide, such as increased homelessness, lack of affordable housing, racial and income inequalities, and obstacles to the accessible and quality education.
- Supporting redevelopment, promoting downtowns, providing resilience public safety services, supporting neighborhood and business districts groups.
- Increasing community engagement, and participation in service delivery.

Environmental Resilience

- Protecting water and other natural resources, rivers, tributaries, streams, lakes, and oceans.
- Reducing pollution, rehabilitate the riparian system to restore the ecosystem to a more natural state.
- Provide a greater access to parks and recreational opportunities to contribute to the overall health and vitality of the community. Decreasing energy consumption, increase renewable energy production.
- Minimizing waste, reuse and re-purposing materials, promoting and improving recycling.
- Seeking operational ways to reduce carbon footprint within an organization and a community.

Good Governance and Resilient Operations

- Approaching government operations from resilient and sustainable practices, aiming for fiscal resilience, transparency, accountability, and accessibility, inclusiveness, and intentionality in outreach to increase opportunities for the marginalized population.
- Strengthening infrastructure, improving building envelopes, and transportation options.

The QBL implementation framework enables local government administrators to position and align projects with specific elements of sustainability and resilience. In attempting to meet the QBL objectives, organizations intentionally design the critical components of the implementation strategies. The following list provides a snapshot overview of some of the key resilience initiatives and strategies, followed by a more in-depth review of sustainable energy and water protection strategies in the proceeding pages.

Implementation of Sustainability and Resilience Initiatives

Resilient Water Systems

The systems designs for water protection and delivery, stormwater systems, and sewer are essential components in sustainability and resilience planning and strategies for local governments. Local elected officials are passionate about the protection of water resource from reducing water waste in the water delivery system to adopting policies of ending the purchase of bottled water for any governmental function, raising awareness about the contrary, environmental, health, and economic consequences of plastic micro-beads to name a few actions. Former Mayor of the City of Grand Rapids, George K. Heartwell, cautioned that much more needed to be done to "protect the treasured water resources in the State of Michigan and around the nation, to enable future generations to have the same access and to be able to enjoy the water assets that we have today" (Alibašić 2013).

In demonstrating a commitment to environmental stewardship, a tool available to local governments is the asset management planning. Delivering specific levels of service may be achieved through efficient and sustainable management of the stormwater/water/parking or any other operating system. Local governments use a proactive long-term planning for asset management to achieve a resilient operating system and sustainable organizations and to ensure the well-being of the community and a healthy environment for the current and future generations. Asset management includes the planning, design, construction, operation, and maintenance of infrastructure in the organization and community. For example, the asset management recommends emerging practices and specific light infrastructure development as well as improvements to the stormwater management to improve the water quality in the river. Those initiatives may include bioswales, cisterns, wetlands, trees, rain gardens, rain barrels, green roofs, permeable pavement, parks and open spaces, green streets and alleys, and other best practices, controlling water on-site.

QBL Categories: *Prosperous and Resilient Economy, Environmental Resilience, Good Governance, and Resilient Operations.*

Resilient Transportation Infrastructure and Systems

Communities require a diversified mode of transportation, accessible public transit, functional roads, and safe traffic conditions. A significant component in the system-wide assessment of the quality of life is the state of the community's facilities, infrastructure, services, and amenities, measured in the amount and quality to meet that community's needs and expectations. For example, a component of the road asset management plan may include the expansion of bike lanes and reduction in the number of traffic lanes in communities.

QBL Categories: *Prosperous and Resilient Economy, Environmental Resilience, Good Governance, and Resilient Operations.*

Resilient Public Safety Services

Local governments are required to provide exemplary public safety services and to have emergency preparedness and disaster mitigation plans in place.

QBL Categories: *Social Equity and Resilient Communities, Good Governance, and Resilient Operations.*

Carbon Footprint Reduction

Each activity identified as operational efficiency and service delivery is directly or indirectly associated with efforts to reduce GHG emissions and decrease carbon footprint.

QBL Categories: *Environmental Resilience, Good Governance, and Resilient Operations.*

Waste Minimization and Recycling

Waste minimization, reduction, recycling, and repurposing of materials are some of the resilient and sustainable initiatives to lessen the negative effects on the environment stemming from production and service delivery.

QBL Categories: *Prosperous and Resilient Economy, Environmental Resilience, Good Governance and Resilient Operations.*

Resilient Energy

Local governments own buildings that consume power and use equipment and vehicles in operations. Administrators are interested in reducing costs of heating and cooling, consumption of fuel, and power consumption.

QBL Categories: *Prosperous and Resilient Economy, Environmental Resilience, Good Governance, and Resilient Operations.*

Resilient Energy System Strategies

Over the past decade, cities have been increasingly active in advocating for environmental responsibility, societal actions to reduce carbon footprint, resilience, and sustainability. For many local governments, energy efficiency has become the cornerstone of their sustainability efforts. Local government administrators actively pursue energy efficiency, energy management, and renewable energy programs and policies, with all three elements defined as a sustainable energy strategy. With local governments' budgets constrained or in crisis, finding cost savings has become an imperative for local governments. In most organizations, energy costs are a notable component of their budgets, and regardless of the geographical position of those local governments, the energy costs are increasing each year.

Most local governments initiated energy efficiency projects, reducing their operational costs, from insulating government buildings, investing in green energy, geothermal; replacing single-pane windows; to upgrading HVAC equipment. Brownsword et al. (2004) model the energy supply and demand for sustainable cities analyzing the "technological and socio-economic aspects of domestic and commercial energy-consumption and use the results to produce a model for urban energy-management" (p. 168). To achieve the most significant energy savings and to integrate the energy efforts into more substantial sustainability and resilience goals, local governments take a more systemic and strategic Quadruple Bottom Line approach to energy efficiency and renewable energy.

The four critical actions for implementing sustainable energy include:

- Institutionalizing energy efficiency and sustainability planning into our organizational culture
- Dedicating staff to data collection and other resources to detailed measurement and progress updates and reporting
- Developing a long-term energy efficiency, energy management, and conservation strategy to guide the energy management work
- Empowering the staff to innovate and pursue smart energy projects

The practical difficulties of changing the infrastructure for energy supply and energy demand on a larger scale come into play inside the built environment. Most state-level policies are not designed to support local communities' implementation of sustainable energy programs, and the role cities play in structural changes of a large-scale shift to sustainable energy infrastructure.

Yergin (2012) attempted to answer the vital question of what the best mix of energy sources is needed to address the energy needs of our world. Moreover, Chen (2011) developed a framework in which it integrates energy demand and supply to link with CO2 emissions in considering "the integration of energy consumption and CO2 emissions" (p. 2695). In the analytical framework for evaluating energy policies, Chen (2011) categorized the environmental issues stemming from energy consumption "into two aspects: exhaustion of natural resources and adverse effects of environmental pollution" (p.2696). The researcher's conclusions focused on three components: energy efficiency improvements, industry restructure, and energy structure improvements. It proposes a theoretical framework with strong corollary connection between energy policies and CO2 emissions.

Resilient Energy Plan and Quadruple Bottom Line (QBL)

Despite the obstacles and decentralized nature of national energy policies, coupled with a diverse patchwork of state-level energy policies, communities across the country are attempting to realize the benefits of sustainable energy programs and projects. The role of the sustainable energy plan for local governments is to identify and prioritize applications based on specific community needs, thereby leading to the implementation of the programs most likely to accomplish those goals. Local governments' focus on sustainable energy efficiency has resulted in projects completed or underway shedding nearly billions of kilowatt hours of electricity from their operations over the past two decades. Even more evident is the communities attempt to capitalize from energy efficiency efforts with the brief but very beneficial availability of the American Recovery and Reinvestment Act (ARRA) funds for Energy Efficiency and Conservation Block Grants (EECBG), which many cities used to produce Energy Efficiency and Conservation Strategy (EECS).

The Energy Efficiency and Conservation Strategy (EECS) serves as a roadmap for becoming a more energy-efficient and sustainable organization and community. Most local governments regularly and consistently analyze, evaluate, and explore cost-effective opportunities for on-site energy generation, including the use of solar panels and geothermal production technologies. The primary focus of the energy efficiency strategy for the local governments was to reduce or avoid cost in operations and improve energy management. In some local governments, with the development of the energy efficiency and conservation strategy, administrators gained a thorough understanding of the carbon footprint for facilities and vehicle fleet. In some instances, organizations used it as an opportunity to collect data and develop a baseline report for all the greenhouse gas emissions generated in the community from residential, commercial, industrial, and transportation-related activities.

Alibašić (2017b) argued the local governments' interest in sustainable energy is to "address budgetary constraints, constituents' demands, sustainability targets, and ongoing changes in energy markets" (p. 4). Along with state governments, municipalities and counties use those strategies as "long-term planning and countering negative economic, environmental, and social aspects of energy production and consumption" (Alibašić 2017b, 4). In reviewing sustainable energy options, and suggesting solutions, MacKay (2009) offered the compelling views on the issues of energy supply, carbon pollution, and the tax system among other items – discussing how and why society is still dependent on certain types of energy sources, by comparing cost of standard or long-established energy sources to a greener or renewable energy sources, such as wind, solar, geothermal, and others.

Analysis of various sustainable energy options, approaches, costs, and benefits is critical for cities and other organizations as they consider projects and policies. Sovacool and Watts (2009) maintained that with the right policy approaches mixed with appropriate leadership, "using today's technology," New Zealand, United States, and the rest of the world can fully be powered by renewable energy (p. 95). The policymakers are introduced to challenges to achieve 100% renewable production. Explaining "the benefits of shifting to small-scale, decentralized technology" and implementation of the feed-in tariffs, paying "renewable energy producers a fixed, premium rate for every kWh of electricity fed into the grid" to the policymakers is essential (Sovacool and Watts 2009, pp. 96–105).

The 100% renewable energy targets have been in place for some time now. However, the number of local governments across the United States committing to the renewable energy goal

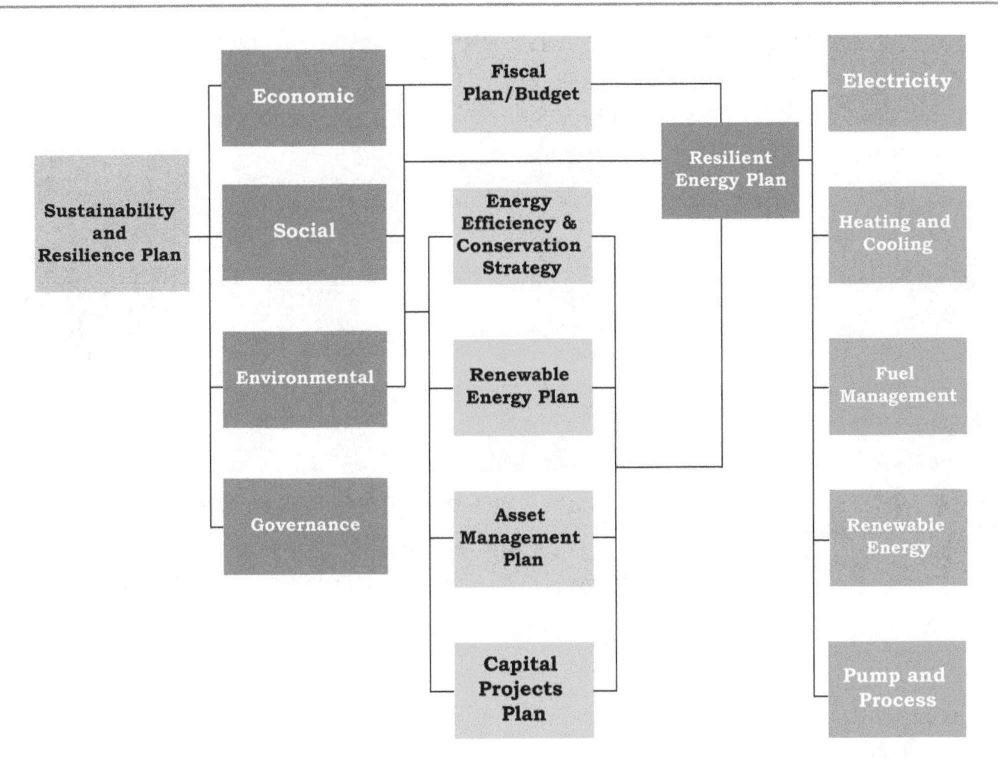

Fig. 5.1 Flowchart – Resilient energy plan and sustainability and resilience planning and implementation

has increased after the US administration's decision to withdraw from the Paris climate agreement (UNFCCC 2015; U.S. State Department 2017). Over 250 US mayors and the city councils/city commissions adopted the 100% renewable energy targets (USCM 2017). The EPA regularly updates and features top 30 organizations from the private and public sector, including local governments procuring or producing renewable energy (EPA 2018). Local governments contemplate the issue of carbon dioxide (CO2) and the impact of human activity on climate change, potential damaging effects on the ecosystem in defining the reasonable policy approaches to energy. However, the perceived ease with which objectives of moving to 100% renewable energy can be achieved is an issue that local governments will have to address in the coming years. Alexander and Boyle (2010) argued for renewable power and evaluated various types of energy sources, and the overall worldwide impact of renewable energy sources "was already providing a significant proportion of the world's primary energy" (p.14). Kohl (2000) posited how renewable energy sources are on the rise, and "often, the potential of the various technologies which exploit renewable energy sources is regarded with skepticism" (p.6). Again, local governments' ability to be in leading position to evaluate, explore and implement the most effective renewable energy technologies is essential.

Realistically and in meeting the expectations of the constituents in their communities, local governments have undertaken the following actions to achieve the goals of sustainable and resilient energy plan and strategies for implementation:

- Develop energy conservation strategies to focus on energy efficiency to decrease energy consumption and demand throughout the organization and in the community.
- Set a goal to achieve an exact percentage of the city's power from renewable sources by a target year to diversify energy sources.
- Reduce total fuel consumption by a certain percentage over the period of time, equating it to the annual savings and carbon footprint reduction (Fig. 5.1).

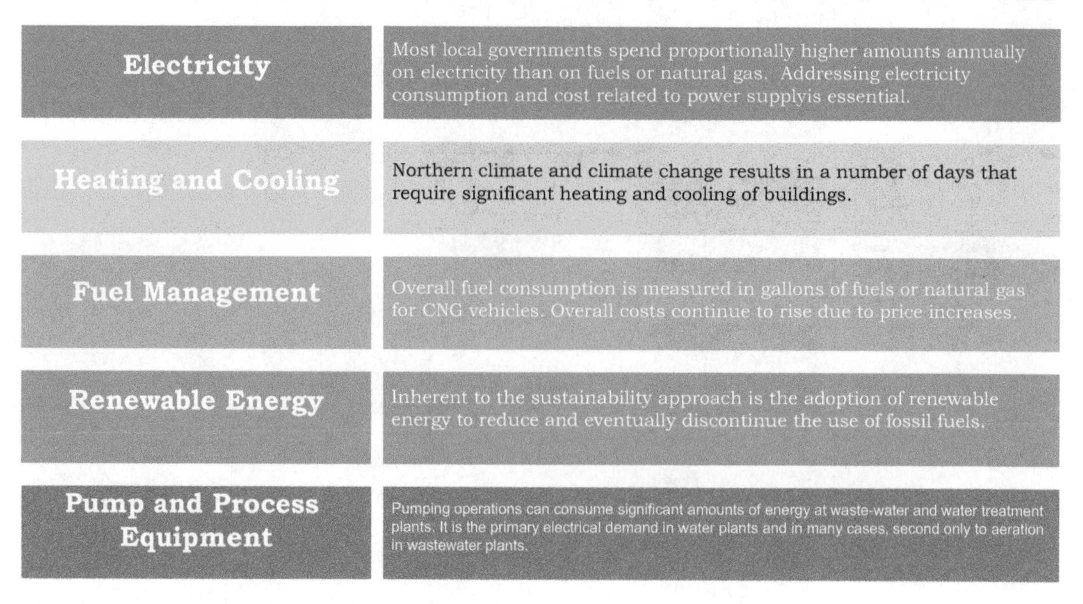

Fig. 5.2 Table – key components of sustainable energy systems

A more in-depth review reveals the specific areas of operations within an organization-wide system, including the pumps and processes in operations, equipment, fuel management, heating and cooling, power production or consumption, and renewables (Fig. 5.2).

Institutionalization of Energy Conservation in Operations

Local governments attempt to ingrain sustainability and resilience strategies into the organizational culture. A range of staff from departments across the city organization and departmental lines are involved in energy efforts, continually seeking energy-saving opportunities. The sustainability and resilience plan is used to promote this institutionalization: the method may be used by each department to plan activities and justify budgets based on the Quadruple Bottom Line pillars. The sustainability and resilience plan holds the department leaders accountable in continuous improvement of service delivery through the enforcement of the "Plan-Do-Check-Act" principles in implementing sustainability and resilience strategies.

Measuring for Success

A commitment to reducing energy consumption is shown through tracking, measuring, and reporting progress, outcomes, and results. Local governments establish an inventory of electricity use and natural gas use for all the city buildings as well as quarterly reporting. By measuring, tracking, and reporting data, city staff identify opportunities for cost-saving measures, and as they began monitoring and reporting on energy efficiency efforts, the city can feature results and updates. Through detailed measurement, at a minimum, and in addition to energy costs, administrators comprehend the carbon effect of muncipal facilities and vehicle fleet. The CO2 equivalent saved by city or other governments in energy efficiency efforts is an environmental benefit to the community.

Empowering Staff and Encouraging Innovation

The energy efficiency strategy enables staff to use innovation and internal know-how to pursue energy projects and then share the emerging practices with each other through internal resilient energy teams set up to examine and recommend future energy improvements and energy-related projects. In most cases, local governments benefit by involving staff to investigate the sustainable energy opportunities and seek innovative solutions to reduce operational cost and help the environment.

Resilient Energy Systems Spotlight: Michigan Cities Building a Resilient Energy Platform

In an era of unremitting economic crisis, local governments explore many venues to lessen the cost and to adopt strategies for financial resilience. Michigan was one of the states hardest hit by the economic recession from 2008. Local governments felt the brunt of the financial disaster. Several cities sought to address revenue losses by building a sustainable and resilient energy platform. As part of strategies to create more sustainable and increasingly resilient communities, the pursuit of a viable energy program is a high priority to cities in Michigan and around the nation. Energy and the associated costs are the common denominators for many communities. While it is evident that comparable platforms have some similarities, there are also significant differences between these communities' approach to sustainable/resilient energy platform. The cities of Holland, Ann Arbor, Grand Rapids, Dearborn, and Farmington Hills are few of those local governments utilizing different energy strategies not only to cut costs but also to increase the renewable energy percentages in their energy portfolio. Common elements to these four cities' strategies are as follows:

- Building awareness and education of the importance of energy to the organization and community, internally and externally.
- Razor sharp focus on energy efficiency and energy conservation.
- Level of renewable energy investments for production or acquisition of renewables in the energy portfolio.

Ultimately, the purpose of the sustainable strategy is to provide sound policy guidance for future energy planning; proper asset management of energy infrastructure, buildings, and facilities; tools for effective management of energy; and transportation needs for the city.

The following provides an overview of these elements for these four cities. Some common factors apply to other cities around the nation and the world.

Resilient Energy in Holland, MI

The Holland Board of Public Works is a municipally owned power utility in the city of Holland, Michigan (HBPW 2017). Having a municipal utility presents a significant opportunity for advancement of the energy goals for the organization and community. In 2011, Holland developed a comprehensive community energy efficiency and conservation strategy establishing a baseline and long-term energy scenarios (Garforth International et al. 2011). The action items in the strategy included a district heating program, Industrial Services – the "Holland Full Utility Service Bundle," a building energy labeling program, and community education and outreach, among others. All the elements of community-wide and organizational outreach are recognized in these plans, as well as laser-like focus on energy efficiency and support for renewable energy.

One of the principle writers of Holland's energy plans and the city's planner, Mark VanderPloeg, commented that having a "municipal utility is a benefit to the community" (Alibašić 2014). As Mark described it, the City of Holland has also completed many what could be considered "the low-hanging fruit" energy efficiency projects, including lighting upgrades and innovative downtown pedestrian lighting retrofits (Alibašić 2014; the City of Holland n.d.). A year later, in 2012, the sustainable return on investment (SROI) study was completed on behalf of the HBPW, "examining the affordability, reliability, social, economic, environmental and health impacts of several generation options," and recommending "a natural gas solution with supplemental Power Purchase Agreements for renewable energy" (P21 2016b).

The city took its sustainable and resilient energy efforts a step further. In 2014, the Holland City Council and the Holland Board of Public Works approved the pricing and sale of $158.84 million in municipal revenue bonds to help fund building a combined cycle natural gas power generating facility (HBPW 2014). The Holland Board of Public Works invested in renewables and significant energy efficiency projects, including downtown heating and cooling. Its commitment to renewable energy through Holland BPW was demonstrated through a 20-year contract with the Michigan Public Power Agency (MPPA) (HBPW 2014). The HBPW built a new natural gas facility, with the excess thermal heat enabling expansion of the existing snowmelt system and providing an opportunity to start the district energy system and heating and cooling system (P21 2016a). Immediately after its completion, the Holland Energy Park received "the Institute for Sustainable Infrastructure's (ISI) Envision Platinum award recognizing the sustainability of public infrastructure," as the "first-in-the-country baseload power generating plant to receive an Envision rating" (P21 2016c).

Resilient Energy in Ann Arbor

The City of Ann Arbor administrators deployed similar strategies to those used in Holland and other Michigan cities. Ann Arbor has a strong reputation for sustainability planning, concentrating on climate change and sustainable energy, in the areas of energy efficiency, energy management, community engagement, and renewable energy. Sustainability is institutionalized in the city's planning process, where communication of the progress in sustainability-related fields is crucial. The city's Sustainable Action Plan contains both quantitative and qualitative targets, and Ann Arbor factors it into its master plan to integrate sustainability and to encourage accountability and transparency.

Ann Arbor has used numerous sustainable energy strategies as part of the sustainability planning. In comparing sustainable energy strategies to other cities in Michigan, Ann Arbor has an active engagement in financing energy performance improvements in commercial buildings using the Property Assessed Clean Energy (PACE) districts to underwrite energy efficiency improvements and renewable energy system installation on commercial and industrial properties through voluntary special assessments bonds (the City of Ann Arbor n.d.-a). In addition to power demand reduction efforts, employing existent data and "forecasting future scenarios," the city commenced a 30% renewable energy goal for municipal operations by 2015 (City of Ann Arbor n.d.-a). Moreover, Ann Arbor through its a2energy outreach and education program is reaching out and promoting energy efficiency and renewable energy to residents and business owners, with tips for energy savings and other resources (City of Ann Arbor n.d.-b). The ability to engage residents and business owners is beneficial to the city in their efforts to affect the energy consumption reduction and reduce the overall carbon footprint in the community.

Resilient Energy Strategies in Dearborn, MI

Other local governments in Michigan have also explored and implemented sustainable energy in their operations. The local government official in Dearborn invested in energy efficiency improvements, solar, and energy performance contracting for lighting in their firehouses. The Dearborn representatives have been active in Michigan Green Communities Challenge, statewide renewable energy and environmental leadership initiatives launched in 2009, and where the City of Dearborn received a gold certificate designation for its efforts in sustainable energy and resilience (Dalbey 2016). David Norwood, the city's sustainability manager, noted the importance of sustainable energy initiatives in meeting the overall sustainability and resilience objectives, as the "necessary effort to reduce energy consumption and engage city staff in seeking innovative solutions for the city's operational efficiency and reduce cost of service delivery for the community" (Norwood David. 2018. email communication with author, January 22, 2018).

Resilient Energy in Farmington Hills

Additionally, Farmington Hills administrators implemented an innovative performance contracting project for efficiency upgrades and conservation measures. Moreover, the city established the Commission for Energy and Environmental Sustainability (CEES) as an advising body to the city council. Initially, the task of the commission was to advise the city on energy savings measures. The scope of the commission task was expanded to engage the community on energy and environmental issues (SustainableFH n.d.). Energy efficiency and energy management strategies are central to the local governments' resilient energy plan. Positive outcomes related to a successful deployment of sustainable energy platforms in Michigan cities have a ripple effect on communities and organizations. There are other resilient energy programs and project opportunities to explore with a potential benefit to cities around the world.

Resilient Energy Strategies in Grand Rapids, MI

The energy conservation efforts in the City of Grand Rapids date back to 1987 when the first facility audits were conducted, implementing basic cost reduction strategies, and evolving into more advanced energy reduction strategies. In 2009, the city developed its Energy Efficiency and Conservation Strategy (EECS) as a blueprint for sustainable energy management. Implementation of the EECS was funded by the Department of Energy's Energy Efficiency and Conservation Block Grant (EECBG). The EECS is a fluid, visioning document, periodically refreshed as technologies evolve and budgets warrant, and provided a framework to select which energy strategies the city would pursue (Alibašić 2012).

In recent years the city has implemented several energy efficiency projects, including the replacement of 40-year-old windows at the City Hall building, light fixture replacement, installation of motion sensors, and other projects. Each city department participates in projects that make their operations more efficient. In addition to monetary value, reduced costs of operations, avoided costs, and savings, projects undertaken by the city contribute to the overall reduction of the city's carbon emissions and carbon sequestration. Some of the more notable energy efficiency projects include the following:

- *Variable-speed water pump at the city's pumping station.* The variable-speed pump allows the city to match the pumping energy to the actual water demand, reducing electricity use and saving the water system money.
- *LEED-certified designation for Wastewater Technical Services Building.* Gained several new energy-efficient upgrades: LED lighting, on-site stormwater retention and treatment, energy-efficient heating, and an improved building envelope.
- *New technology to monitor treated wastewater.* The new system allows for more efficient operation of treatment equipment, reducing electricity use. In addition to savings, the City received a $57,000 energy efficiency incentive payment for this upgrade.

Furthermore, efforts to reduce energy consumption continued to pay annual dividends toward social and environmental and governance goals and targets. Since FY 2009, the city has reduced electricity usage by 5.53%, a reduction of over 5,800,000 kWh, and has continuously implemented energy efficiency projects on city-owned buildings which have helped lower the overall consumption of energy. However, as with virtually all other goods, the price of electricity has gradually risen over the years. This increase in energy prices has created a net increase in

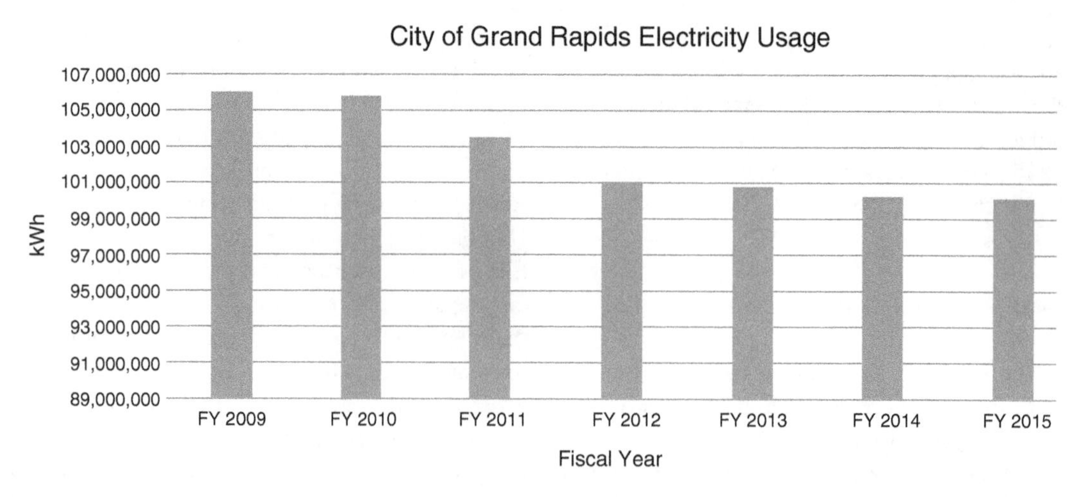

Fig. 5.3 Chart – City of Grand Rapids energy consumption FY 2009–2015

energy expenses for the organization, even with lower overall energy consumption. In FY 2009, the use of electricity in city-owned buildings totaled approximately 106 million kWh. The city's electrical consumption is at an annual level below 103 million kWh, with the use in FY 2015 reaching the lowest point of 100.1 million kWh. Over the same period, the energy cost increased on average 6% annually. Energy efficiency efforts are in line with targets in the sustainability plan, and energy and cost savings support the investments. In the most recent version of the city's sustainability plan, one of the energy targets is to "increase cost avoidance due to energy inefficiency by an additional 3% over FY15 results in City facilities by June 30, 2021" (City of Grand Rapids 2016, p. 23, Fig. 5.3).

Local governments achieve further electricity savings with the use of grants, energy optimization rebates, savings, and one-time investments.

Encouraging and Empowering Innovation Among Staff

The city organized the sustainable energy team, which consists of staff from various departments involved in energy management decisions relative to energy use in city-owned facilities, fleets, or equipment. All types of energy uses are scrutinized, including natural gas, steam, electricity, and fuel. The team incorporated a broad cross section of staff from across city departments.

In line with transformation investment strategy, staff from Water, Environmental Services, Parking, Facilities, Parks & Recreation, Fire, and Economic Development departments joined to develop a resilient energy strategy to serve as a blueprint document to be regularly updated as technologies evolve and budgets warrant. It presented opportunities to select plans to consume energy more efficiently, reduce greenhouse gas emissions, bring down energy and fuel use, lower energy costs, and support efforts to meet the renewable energy target. As the city does not own its electric utility, it is imperative to work in partnership with the private sector to meet the renewable energy target.

City staff are involved, seeking resilient energy strategies to lessen the costs and improve the environment, and leading to innovative solutions. For instance, former Technical Control Supervisor, Laron Morgan, devised a Heat Recovery Program at the city's wastewater plant to conserve on natural gas usage. The project involved transferring heat from a facility where the process produces too much heat to an adjacent building during winter months. Heating the adjoining building with natural gas used to cost $20,000 per year, but with sufficient waste heat to thoroughly heat the building without using natural gas. The project cost was roughly $100,000, with the expected payback on the investment in less than 5 years. The project won the 2012 DTE Energy Optimization Award. An organizational focus on resilient energy spurs innovation, new ideas, and ongoing commitment from all city staff in every department.

Resilient Community Energy

Externally, city concentrated the efforts on designing a resilient energy platform for energy audits in neighborhoods and tools to increase energy performance in homes. The city partnered with the West Michigan Environmental Action Council (WMEAC) using external grants to complete energy audits in neighborhoods through the Better Building for Michigan program. Eventually, almost two thousand residential homes had energy appraisals performed and the energy efficiency improvements made in households participating in the program.

Resilient Renewable Energy Strategy

In late 2007, after negotiating better terms and conditions and price per kWh with the power utility, the city entered into a partnership arrangement with the power utility Consumers Energy to procure 20% of its energy from renewable sources. Additionally, after the installation of the solar photovoltaic system at the Water Administration building, geothermal at two fire stations, and commitment through enterprise system to procure green energy purchases, the City achieved green power targets as a percentage of total electricity use. Grand Rapids is featured frequently on the EPA's top 30 local government list, which represents the most significant green power users among local government partners in the Green Power Partnership (EPA 2018). The city leadership furthered its commitment to green energy by establishing a 100% renewable energy target by 2025. To meet the renewable energy target, city administrators used an opportunity to reduce energy consumption and then use the cost avoided and savings to reinvest into renewable energy projects and purchase.

The City's 125 KW, photovoltaic power generation system, is using solar panels on the LEED-certified Water Service Facility. The system was operational in June of 2012 and is offsetting over 35% of the facility's annual electric consumption, producing nearly 638,000 KWh of green energy since installation in June of 2012 and offsetting close to 383 tons of carbon dioxide emissions (City of Grand Rapids Water Facility 2018). The City has been evaluating other renewable energy projects, including a potential large-scale wind project at its water filtration plant, a solar project at a former landfill site, a solar project at its water filtration plant, and a large-scale bio-digester project. After two setbacks with the initial phases of wind and solar projects, and lessons learned from those two projects, the city broke ground on its bio-digester project and is still actively pursuing the 100% renewable energy target (Alibašić 2015; Balaskovitz 2017; City of Grand Rapids 2010; Holland Sentinel 2009; Huffman 2017; Steiner 2017).

Resilient Water Service System Spotlight

The City of Grand Rapids provides water, wastewater, stormwater, and other public services regionally to several municipalities across two counties, serving a population of over 280,000 and covering a service area of 137 square miles. The city's ability to deliver water services in a timely, cost-effective, accountable, and socially and environmentally conscious way are the most critical components of sustainable water service delivery to residents and businesses. The city carefully aligns its sustainability plan with targets and outcomes to protect water resources and deliver water services. The inclusion of water resource resiliency targets in the city's yearly sustainability plan is essential to meeting and implementing those resilience objectives.

- Meeting or exceeding 100% of federal and state drinking water standards with no violations
- Increasing reuse of captured water and "gray water"
- Reducing the annual customer consumption of water provided by the city's water system by an additional 3% over a period

The staff at the water system have been tracking consumption of water overall and on a per meter basis since the year 2000. Over this time, overall use in the system has decreased a total of 16.66% or an average of 1.39% per year, with average consumption per meter dropping by an average of 2.14% per year (Alibašić 2013). Additionally, the city participated in the annual Great Lakes-Saint Lawrence Cities Initiative challenge to its members to reduce their water consumption by 15%. Grand Rapids has reduced its overall water consumption by 8.5 billion liters per year (2,25 billion gallons), exceeding the objective of a 15% reduction before 2015, using the 2000 water consumption as a benchmark. Water conservation was accomplished through the implementation of a series of emerging practices, including a loss reduction program in the city's water distribution system, public awareness campaigns using different communication methods, the use of reused and raw water for irrigation purposes, and changes in the plumbing practices (Fig. 5.4).

Resilient Stormwater Management Spotlight

Recently, the City of Grand Rapids brought to completion a 27-year-long project, eliminating "the combined storm and sanitary sewer system beneath the oldest portions of the city" and removing all of the fifty-nine combined sewer overflow (CSO) points (Wilcox 2015). The receiving body of water for combined sewage overflows is the Grand River, a tributary of Lake Michigan and one of the largest rivers in the basin. In the last two decades, the City of Grand Rapids embarked on a comprehensive program to eliminate all combined sewer overflows (CSO) in the city. By making these long-term, strategic investments in infrastructure of almost $300 million to separate sanitary and stormwater systems,

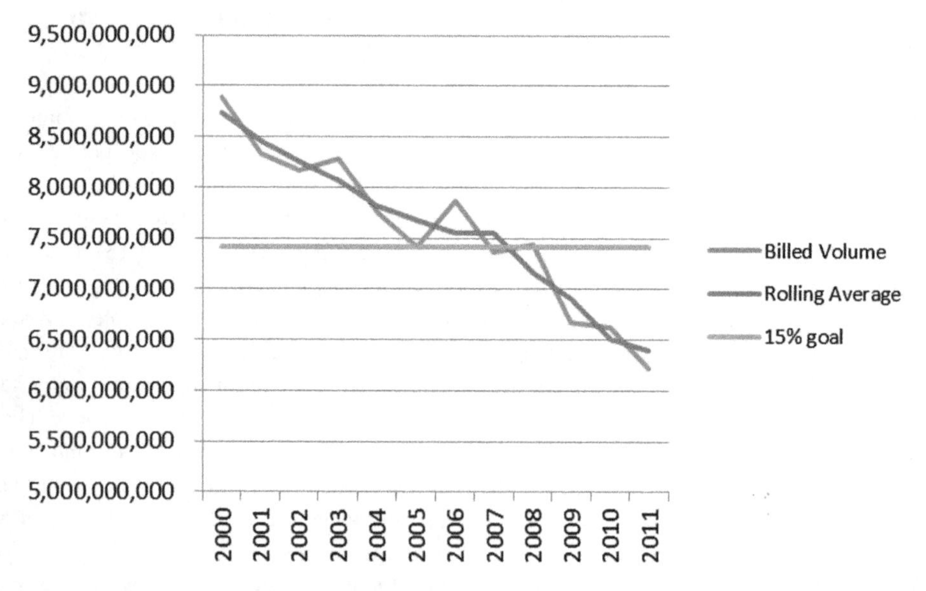

Fig. 5.4 Chart – water consumption per year in Grand Rapids. (Source: Water Services Department – 15 Percent Goal Worksheet "Rolling averages" for Billed Volume)

the local government completely removed all the CSOs points and had significantly reduced pollution to the Grand River.

The city's wastewater treatment plant is a success story in the reuse of its treated effluent for plant operations. Some work remains for the local government administrators to provide critical stormwater management services. As stated by Mike Lunn, Grand Rapids' Environmental Services Manager, "the city's approach to projects is first to evaluate green infrastructure potentials and incorporate them before designing the traditional gray infrastructure" (Alibašić 2013). Green and natural stormwater management systems along with environmental, governance, and social benefits deliver positive economic results too. It is also less expensive to maintain them in comparison to the traditional gray infrastructure. Goals of this new approach are the paradigm shift in stormwater management that includes proactive planning and an improved technical method of doing business.

Partnership for Resilience Prosperous programs involve innovative and pragmatic alliances to offset and leverage resources in the world fraught with financial instabilities. Financing sustainability and resilience-related projects requires the engagement of banking sector, the private sector, and public sectors in ways not seen in the past. One of the critical patterns in the adoption of sustainability and resilience is the jump from planning to implementation. The partnership between local governments and private sectors can facilitate such jumpstart. The municipal government and the community are fully committed to the principles of sustainability and resiliency support the natural environment, economic system, good governance, and social infrastructure, the core elements and principles of the Quadruple Bottom Line. Steadfastness toward a more sustainable and resilient future is evidenced in collaboration and partnership throughout the community, through sustainability partnerships, regional climate action plans, and other programs to share emerging practices, to leverage resources, and to transform regions as the centers of sustainability and resilience.

Summary

As local governments continue to increase the resilience of their water, parking, building, fleets, energy, and other systems, they continue to adopt new policies and strategies that are consistent with the Quadruple Bottom Line. Contemporary cities, townships, villages, and counties approve multi-year sustainability and resilience plans and strategies to implement comprehensive, measurable sustainability and resilience initiatives and projects. Implementing projects reduces the cost and decreases carbon footprint and dependence on and consumption of fossil fuels. The goal is to remain at the leading edge of resilient community practices, saving taxpayers money and creating a more vibrant, healthy environment for residents. There is evidently still a wealth of new technologies and cost-saving opportunities to tap into for municipalities. Local governments implement sustainability and resilience initiatives as the most effective long-term strategy in meeting environmental, social, economic, and governance objectives for an organization and a community at large.

Energy is a substantial cost driver for a myriad of organizations. Local governments strive to ensure savings in their operations by efficiently addressing energy costs. Notably, increased electricity costs are a budgetary burden for organizations. Additionally, decreased renewable energy costs have made a return on investment for renewable energy more economically and financially feasible. To achieve the most energy savings and to integrate the sustainable energy work into greater resilience goals, local governments take a more systemic, holistic, and strategic Quadruple Bottom Line approach to energy planning and initiatives implementation. Local governments build a significant level of awareness using internal resources throughout the organization to promote energy efficiency and renewable energy. The sustainable energy strategies are being developed on the principles of reducing electricity and natural gas consumption and costs, reducing transportation-related fuel consumption, and meeting renewable energy and greenhouse gas emissions targets.

Further Discussions

- Assess the implementation strategies for sustainability and resilience projects.
- Analyze the impact of sustainable energy strategies in organizations.
- Discuss the types of sustainability and resilience projects.
- Evaluate the implementation strategies and QBL approach to implementation.
- Address the issues of water quality protection.

References

Alexander G, Boyle G (2010) Introducing renewable energy. In: Boyle G (ed) Renewable energy: power for a sustainable future, 2nd edn. Indian Edition, Oxford

Alibašić H (2012) How energy efficiency strategy pays off in Grand Rapids. TriplePundit People, Planet, Profit. Retrieved from http://www.triplepundit.com/2012/08/energy-efficiency-strategy-pays-grand- rapids/

Alibašić H (2013) Building sustainable and resilient Grand Rapids. German American Water Technology Magazine 2013/2014. German American Water Technology Initiative

Alibašić H (2014) Michigan cities building a sustainable energy platform: Holland, Ann Arbor and beyond. TriplePundit People, Planet, Profit. Retrieved from https://www.triplepundit.com/2014/05/michigan-cities-building-sustainable-energy-platform-holland-ann-arbor-beyond/

Alibašić H (2017a) Measuring the sustainability impact in local governments using the Quadruple Bottom Line. Int J Sustain Policy Prac 13(3):37–45

Alibašić H (2017b) Sustainable energy policy for local and state governments. In: Farazmand A (ed) Global encyclopedia of public administration, public policy, and governance. Springer International Publishing AG, Cham

Balaskovitz A (2017) Michigan solar project scrapped after developer 'disappeared,' city claims. Midwest Energy News. Retrieved from http://midwestenergynews.com/2017/01/25/michigan-solar-project-scrapped-after-developer-disappeared-city-claims/

Brown LR (2006) Plan B 2.0: rescuing a planet under stress and a civilization in trouble (updated and expanded). Earth Policy Institute, New York

Brownsword RA, Fleming PD, Powell JC, Pearsall N (2004) Sustainable cities– modeling urban energy supply and demand. Appl Energy 82:167–180. https://doi.org/10.1016/j.apenergy.2004.10.005

Chen C (2011) An analytical framework for energy policy evaluation. Renew Energy 36(10):2694–2702. https://doi.org/10.1016/j.renene.2011.02.023

City of Ann Arbor (n.d.-a) Property Assessed Clean Energy (PACE). Retrieved from https://www.a2gov.org/a2energy/commercial/Pages/faqs.aspx

City of Ann Arbor (n.d.-b) a2energy. Retrieved from https://www.a2gov.org/a2energy/Pages/default.aspx

City of Grand Rapids (2010) Response to questions/Comments about proposed Wind Energy Turbines project (WET) at Lake Michigan Filtration Plan (LMFP). Retrieved from http://www.grand-rapids.mi.us/enterprise-services/Documents/12312_2010-01-29%20Response%20to%20Questions%20from%20GHT%20_FINAL_.pdf

City of Grand Rapids (2016) Sustainability Plan FY2017-FY2021. Office of Energy and Sustainability. Alibašić H (ed). Retrieved from https://www.grandrapidsmi.gov/files/assets/public/departments/office-of-sustainability/reports-and-documents/office-of-sustainability/sustainability-plan-fy17-fy21.pdf

City of Grand Rapids Water Facility (2018) SolrenView. Retrieved from https://www.solrenview.com/SolrenView/mainFr.php?siteId=1341

City of Holland (n.d.) Downtown pedestrian lighting: LED Retrofit Project. Retrieved from http://www.cityofholland.com/sites/default/files/fileattachments/downtown_led_brochure_2011_0.pdf

City of Holland (2012) Special City Council Study Session with the Board of Public Works Agenda. Retrieved from http://www.mlui.org/userfiles/filemanager/1455/

Dalbey B (2016) Dearborn wins gold for energy, environmental initiative. Patch. Retrieved from https://patch.com/michigan/dearborn/dearborn-wins-gold-energy-environmental-initiative

Garforth International, LLC, City of Holland, Holland Board of Public Works (2011) Holland community energy efficiency and conservation strategy: creating a global competitive community. Project Work Team Report. Retrieved from: http://www.cityofholland.com/sites/default/files/fileattachments/final_cep_for_suscom_sept_9_2011_for_website.pdf

Hardjono T, Van Marewijk MV, de Klein PD (2004) Introduction on the European Corporate Sustainability Framework (ECSF). J Bus Ethics 55(2):99–113. https://doi.org/10.1007/s10551-004-1894-x

Holland Board of Public Works (HBPW) (2014) Holland approves sale of bonds to finance new power plan. Retrieved from http://p21decision.com/2014/03/20/holland-approves-sale-of-bonds-to-finance-new-power-plant/

Holland Board of Public Works (2017) About. Retrieved from https://www.hollandbpw.com/

Holland Sentinel Editorial Board (2009) Our view- Siting wind turbines: yes, in our backyard. Retrieved from http://www.hollandsentinel.com/article/20091202/news/312029798?start=2

Huffman B (2017) Grand Rapids breaks ground on $38 million biodigester sludge treatment center. Michigan Radio. Retrieved from http://michiganradio.org/post/

grand-rapids-breaks-ground-38-million-biodigester-sludge-treatment-center

Kohl H (2000) Renewable energy sources on the rise. In: Wengenmayr R, Bührke T (eds) Renewable energy: sustainable energy concepts for the future. Wiley-VCH, Weinheim, pp 4–14

MacKay DJC (2009) Sustainable energy: without the hot air. UIT Cambridge Ltd, Cambridge, UK

Power for the 21st Century. (P21) (2016a) P21: Looking back. Retrieved from http://p21decision.com/2016/05/05/p21-looking-back/

Power for the 21st Century. (P21) (2016b) How the Energy Park will improve snowmelt. Retrieved from http://p21decision.com/2016/01/29/how-the-energy-park-will-improve-snowmelt/

Power for the 21st Century. (P21) (2016c) Holland Energy Park earns Envision® Platinum Award. Retrieved from http://p21decision.com/2016/07/20/holland-energy-park-earns-envision-platinum-award/

Sovacool BK, Watts C (2009) Going completely renewable: is it possible (let alone desirable)? Electr J 22(4):95–111. https://doi.org/10.1016/j.tej.2009.03.011

Steiner A (2017) The road to 100: Grand Rapids' journey to be Michigan's first all-renewable-powered city. Rapid Growth. Retrieved from http://www.rapidgrowthmedia.com/features/GRrenewable100.aspx

Stubbs W, Cocklin C (2008) Conceptualizing a "sustainability business model". Organ Environ 21(2):103–127. https://doi.org/10.1177/1086026608318042

Sustainable Farmington Hills (SustainableFH) (n.d.) About CEES. Retrieved from http://www.sustainablefh.com/Menu/About-CEES.aspx

United States Environmental Protection Agency (EPA) (2018) Green power partnership. Green power partnership top partner rankings. Retrieved from: https://www.epa.gov/greenpower/green-power-partnership-top-partner-rankings

U.S. Department of State (2017) Communication regarding intent to withdraw from Paris Agreement. Office of the Spokesperson. Retrieved from: https://www.state.gov/r/pa/prs/ps/2017/08/273050.htm

United States Conference of Mayors (USCM) (2017) 85th Annual Meeting 2017 adopted resolutions. Retrieved from http://legacy.usmayors.org/resolutions/85th_Conference/proposedcommittee.asp?committee=Energy

United Nations Framework Convention on Climate Change (UNFCCC) (2015) Adoption of the Paris Agreement. United Nations framework convention on climate change. Retrieved from. https://unfccc.int/resource/docs/2015/cop21/eng/l09r01.pdf

Wilcox K (2015) Grand Rapids eliminates Combined Sewer Overflows (CSO), adds green spaces. Civil Engineering. The Magazine of the American Society of Civil Engineers. Retrieved from http://www.asce.org/magazine/20150922-grand-rapids-eliminates-combined-sewer-overflows,-adds-green-spaces/

Yergin D (2012) The quest: energy, security, and the remaking of the modern world. Penguin Press, New York

6 Examining the Intersection of Sustainability and Resilience

"But, whatever our resources of primary energy may be in the future, we must, to be rational, obtain it without consumption of any material. Long ago I came to this conclusion, and to arrive at this result only two ways, as before indicated, appeared possible- either to turn to use the energy of the sun stored in the ambient medium, or to transmit, through the medium, the sun's energy to distant places from some locality where it was obtainable without consumption of material." Page 199 of Tesla, N. (1900). The problem of increasing human energy: With special reference to the harnessing of the sun's energy. The Century Magazine

Key Questions

The goal of the sixth chapter of this book is to answer the following underlying assumptions and questions:

- What are the climate resilience, resilience, climate preparedness, climate mitigation, and climate adaptation strategies?
- How is resilience planning connected to sustainability planning within organizations?
- What strategies can local governments utilize to integrate climate resilience into organizational planning, including short- and long-term objectives?
- How do local governments plan for climate change threats?
- What are the appropriate steps to integrate climate change strategies into emergency preparedness and disaster mitigation?
- What do communities need to look at in addressing the impact of climate change?
- What are the economic effects of climate change?

Introduction

Local government administrators face daily challenges as they manage the provision of services to residents and businesses in their communities. The new global and regional realities of climate change and extreme weather are going to affect localities and most importantly local governments in their provision of an array of critical services, such as public safety, infrastructure, water, and waste management to name a few. Local units of governments are attempting to address the vital vulnerabilities of communities to the climate change, the issue of greenhouse gas emissions, and symptoms related to an increased environmental pollution. In recent years, climate change generated additional challenges for local government officials exemplified through extreme and unpredictable weather patterns, including but not limited to heat waves, intense rain events, more frequent flooding, changes in temperatures, snowstorms, hurricanes, and droughts. While the impacts of the climate change are region specific and are diverse in intensity and the impact levels, they affect all facets of healthy communities.

Most notable impacts include water quality and freshwater resources, power outages and disruptions as demand for cooling increases during heat wave events, demands on first responders, increased stressors on infrastructure, economy, service delivery, and vulnerable population. Local government leaders plan for and imple-

H. Alibašić, *Sustainability and Resilience Planning for Local Governments*, Sustainable Development Goals Series, https://doi.org/10.1007/978-3-319-72568-0_6

ment robust climate mitigation, adaptation, and climate preparedness strategies to ensure maximum community and organizational resilience. Cities are the most vulnerable to climate change, and the implementation of climate adaptation and mitigation strategies is most effective on a localized scale.

Some of those measures may include:

- Developing energy conservation and efficiency strategies to reduce energy consumption and demand throughout the organization and plan for peak load demands in collaboration with power utilities.
- Setting a 100% renewable energy target for the city's operations by target year. Diversified energy sources and decentralized power delivery are essential for local resilience and greenhouse gas reduction.
- Reducing total fuel consumption in fleet and operations.
- Setting a goal to increase the tree canopy cover and to diversify the type of tree species planted to increase resilience to urban heat island effects and heat waves.
- Effectively managing waste minimization, reduction, and recycling of materials.
- Providing exceptional public safety services and developing and implementing emergency and disaster preparedness plans and strategies.
- Reducing water consumption and protect water and other natural resources.
- Ensuring economic development, planning, and engineering services are provided in the system-wide, holistic approach, in partnership with local and regional business and economic development agencies and other institutions for maximum outcomes.

As noted in the diagram, in recognizing the completeness and complexities of the systems, the elements of climate resilience are implemented through existing sustainability and climate action plans (Fig. 6.1).

It is compelling for local governments to adopt and undertake various approaches to mitigate and adapt to the impact of climate change. Hallegatte et al. (2011) suggested "climate change mitigation strategies may also lead to a diversification of energy sources, which in turn would decrease systemic losses due to a disruption of supply (not necessarily due to climate change)" (p. 80). A starting point to climate resilience planning is the completion of the regional or local resilience report, with a specific and detailed understanding of climate change data, weather patterns, and localization of the climate change impact.

Defining Resilience

Resilience may be characterized as "an attempt to prepare for the worst and to be able to rebuild from disaster," and "in the context of effective strategies undertaken by communities to prepare for unforeseen and unpredicted events as a result of climate change and extreme weather events, and their ability to revive after the disaster in a sustainable manner" (Alibašić 2014). The consequences of not planning and adequately preparing for potential disasters can be devastating for human resources, services, buildings, and infrastructure. Fiksel (2003) explained the system resilience in light of "significant disruptions or discontinuities" shifting "the system away from its current equilibrium state" (p. 5333). Fiscal constraints and the impact of a global economy on local governance and the ability to deliver outcomes may be viewed as significant disruptions. Local governments engaged in sustainability and resilience planning can adapt and transform and accept discontinuities as they continue to provide services without interruptions. In a crisis, sustainability-related efforts become an opportunity and a tool for local governments as then they need to change their priorities, reporting, measuring, and outcomes of the budget process.

Preferably to waiting for national leadership on climate change, cities, villages, townships, counties, and other localities take proactive climate preparedness actions in pursuit of the interest of their constituents and residents. Cities have a unique role to provide services and to decide on their own what policy options best

Fig. 6.1 Diagram – resilience and sustainability initiatives

fit the organizational and community framework. In addition to vulnerabilities to climate change and extreme weather event, cities are some of the most prominent contributors to carbon pollution. As Fitzgerald et al. (2012) suggested the "emissions of greenhouse gases leading to climate change, represents the most important current environmental challenge" (p. 371). As such, local government administrators should strive to reduce the carbon impact on the society. Byrne et al. (2006) argued that cities of industrial and more developed nations play an important role in addressing the negative consequences of pollution, specifically in attempting to cease "the currently destructive relationship between urban industrial society and the global environment" (p. 87). The evidence of impacts and consequences of climate change on the environment and societies is global in scope.

Climate Change Concerns for Organizations and Communities

An extensive body of scientific research points to the undeniable and indisputable evidence of the harmful effects of the industrial activities on climate, causing environmental and social disruptions. Extreme weather events, increases in global temperatures, sea level rise, economic disruption, infrastructure damage, species extinction, and weather pattern changes are some of the ongoing and well-documented concerns and challenges for communities, population, and the society as a result of the changes in the climate (Fletcher 2013; Hallegatte et al. 2011; IPCC 2014; Karl et al. 2009; Lindfield 2010; Mach et al. 2016; Malcolm et al. 2006; Pecl et al. 2017; Segan et al. 2015; Stott et al. 2016; The World Bank 2012; Urban 2015; USGRCRP 2017; Visser et al. 2014). Dong et al. (2014) and Silva et al. (2013) examined the linkages between the human-produced carbon emissions, climate change effects, and mortality rate. Beyond the sea level rise and acidification of the oceans, the threats to the most extensive body of freshwater, Great Lakes, are well documented (Kling et al. 2003).

The global community has been hesitantly reacting to the threats of climate and changes. The failure of national governments to address climate prompted leadership actions on subnational level, mainly by local governments. With direct consequences forecasted, cities are at the forefront of those efforts to effusively deal with climate change.

Community-Wide Resilience

Community-wide resilience preparedness considers emergency preparedness, energy planning, health, human resources, and public safety issues. Besides, the "chief features of climate change actions at the local governments' level are the cities' ability to prepare their respective communities to be more agile and adaptive to extreme weather events and disasters" (Alibašić 2018a, p.4). The benefits of addressing and taking the climate change trends into consideration far outweigh the costs associated for climate resilience and preparedness planning. Furthermore, "the resiliency to crisis and disaster is the capacity and adaptability of systems not only to withstand stresses and shocks but also to continue to thrive during and after the disaster" (Alibašić 2018b, p.1). Weather patterns cannot any longer be predicated upon existing models, and the impact on regions, cities, and especially urban areas are already immense. Besides, the more frequent sever weather events and changes in precipitation and temperature patterns impede the social system, governing, ecosystem, and the economy. Moreover, the major climate trends such as air and water temperature variations and increases, droughts and dry seasons, the frequency and intensity of storms, and floods impact the spectrum of sectors.

Climate Change and Emergency Preparedness

Climate change represents a whole set of extra challenges in emergency planning, preparedness, and disaster mitigation for municipalities. At a minimum, emergency plans incorporate the latest science to understand impacts of such changes better and develop various alternatives. A changing climate generates many challenges for state, local, and tribal governments as elected leaders, planners, and resource managers consider mechanisms for ensuring community resilience and preparedness.

In response to heat wave events, cities made adjustments to their emergency action guidelines to coordinate services with the nonprofit agencies and utilities, in regard to resources, facilities, and cooling centers. Local governments respond more holistically to heat waves and hazardous rain events and have more control over the events arising from climate change and extreme events by tying the sustainability and resilience plan directly to emergency planning. Having the accurate climate data, information about the history of weather patterns and events and infrastructure are critical components of resilience planning (Fig. 6.2).

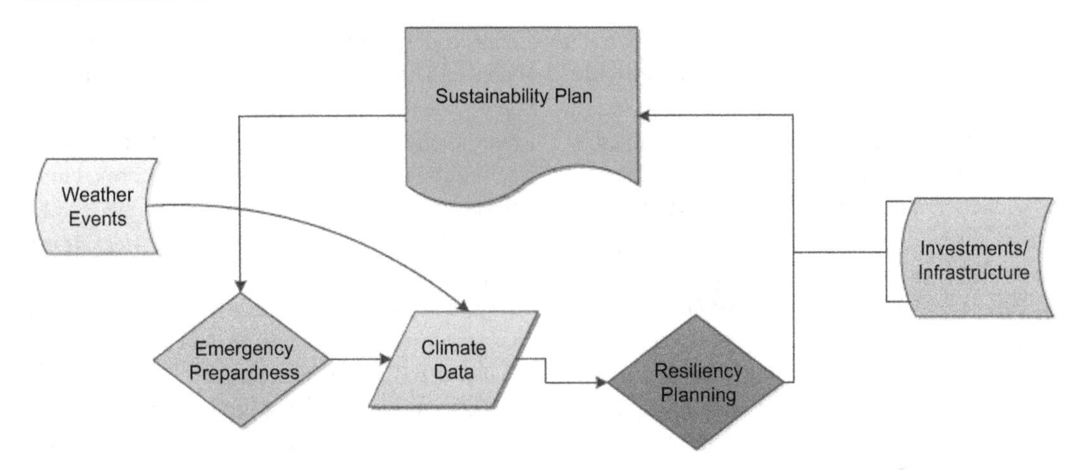

Fig. 6.2 Combining sustainability and resilience planning with emergency preparedness

National Versus Subnational Climate Resilience Policies

A lack of national resilience policies and programs enables decentralized approaches to climate resilience planning, bringing to light the relevance of local governments and their policies toward sustainability, resilience, and climate preparedness. In discussing scaling climate adaptation strategies in urban settings, Brugmann (2012) defined resilience as "the ability of an urban asset, location and system to provide predictable performance" (p. 217). In examining three municipalities in Canada, Burch (2010) reviewed the "varying levels of success at reducing greenhouse gas emissions and enhancing resiliency" in those localities (p. 7575). In the United States, in the period prior and post President Obama's administration, the policies to address the threats of climate change have been nonexistent. During the first decade of the twenty-first century, the US Conference of Mayors established the Climate Protection Agreement with an overarching goal to encourage cities to deploy climate mitigation strategies and reduce the organizational carbon footprint of city organizations. The commitment to reduce carbon footprint was endorsed and actively pursued by over one thousand US mayors, leading to increased leadership by local governments in the United States in the absence of US federal leadership (USCM 2005).

Under President Obama, local and state leaders had a prominent role in planning for the climate preparedness, working directly with the White House staff. In 2013, by Executive Order 136531, Preparing the United States for the Impacts of Climate Change, President Obama established the State, Local, and Tribal Leaders Task Force on Climate Preparedness and Resilience (Task Force 2014). The task force members were asked to examine all venues of responses and preparedness to address climate change trends and threats and recommend climate strategies for federal government to better support local and state actions to make communities more resilient. Additionally, task force members reached out to thousands of government organizations, universities, and other stakeholders, seeking recommendations, focusing on scientific and policy approaches to climate preparedness and resilience strategies (Task Force 2014).

However, under the current administration, not only did the United States announce its intent to withdraw from Paris Climate Accord, but other environmental protections and climate change work are being rolled back under the assault from the current administration (Davenport 2017; United State Department 2017). Many local government and some state leaders in the US declared their intent to continue fulfilling pledges of the international climate agreement. Local and state governments continue to pursue climate preparedness and resilience strategies. As observed by Gordon (2016), cities have undertaken a myriad of actions in addressing the climate change-related threats. The scaling of the climate resilience efforts to local government levels is underscored in the ability to measure and track the progress of outcomes from those activities. Many initiatives by local governments and in some states in the United States are aimed at reducing the GHG emissions, mitigating the impact of climate change or adapting to new climate realities. Recently, 364 US mayors agreed to cut GHG emissions per Paris Climate Accord requirements (Climate Mayors 2017).

Furthermore, over 250 cities adopted the 100% renewable energy targets. Moreover, the EPA top 30 lists are regularly updated, showing significant renewable energy commitments from local governments and other organizations (USCM 2017; EPA 2018). The 100% renewable energy targets are intended to enable cities to mitigate and reduce impacts from energy consumption in city operations by switching energy demand to greener energy. These efforts within and outside organizations engaged in climate resilience and preparedness are necessary to prepare for the next natural disaster that will inescapably occur. Constituents at a local level, cities and communities, expect their appointed and elected officials to meet their challenges and objectives to address concerns and prepare for the climate change-related impacts.

Local Government Roles Reinvented

As cities take a more active role in climate preparedness and resilience, the local zoning and land-use policies are reinvented; new approaches to the procurement of energy and power supply and production procure are adopted. By taking a more proactive role and in some cases partnering with the private sector or academic institutions, local units of government are less dependent on federal and state funding. The reinvented role of local governments and the interaction on subnational, national, and international scale are critical to creating synergies in concerted attempts to reduce, minimize, and ultimately remove the adverse impacts of climate change. The collective body of local government policies, projects, and programs leaves an impressive impact on the environment, society, economy, and good governance. Local leaders and administrators draw from the shared experience of cities around the world leading to better governance of local resources. Environmental, economic, social, and governance issues through the climate change lens are no longer viewed as localized issues. Cities continue to pursue resilience policies, perhaps best described through the combination of sustainability-related efforts copulated with the impacts of climate change and weather events as the combination of resistance and survival strategies.

The local governments' cutting-edge planning for climate resilience in the United States is evident in cities like Austin, TX; Baltimore, MD; Boston, MA; Chicago, IL; Dubuque, IA; Eugene, OR; city and county of Los Angeles and San Francisco in California; Portland, OR; Philadelphia, PA; New Orleans, LA; Broward, Miami-Dade, Monroe, and Palm Beach Counties in Florida; and New York City, NY to name a few (City of Austin 2014 & 2015; Baltimore Office of Sustainability 2013; Broward County 2015; City of Boston 2014; City of Chicago 2008; City of Dubuque 2013 & 2017; City of Eugene n.d.; City of Los Angeles 2015a & 2015b; County of Los Angeles 2015; City of New Orleans 2017; City of New York 2017; City of Philadelphia 2016 & 2017; City of Portland and Mulnomanh County 2017; City and County of San Francisco 2013 & 2017; Southeast Florida Regional Climate Change Compact Counties 2012). Other cities in the US and Canada and around the world are making significant inroads and progress toward climate resilience planning. For instance, the Pensacola City Council in Florida appointed the task force on climate adaptation and mitigation to recommend climate resilience strategies to the council members to aid climate change in the city and region (Baucum 2017). In Arkansas, the City of Fayetteville elected officials published a document entitled Arkansans Can Take Steps to Respond to Climate Change as a call to the state residents to be proactive in combating the threats of a changing climate (City of Fayetteville n.d.). The table lists some of most operational climate resilience plans, climate preparedness, and climate mitigation and adaptation action strategies, with a sample of related programs and departments overseeing them (Table 6.1).

Table 6.1 A sample of local governments with operational climate resilience plans and strategies

Local government	State	Year of release	Title of the plan	Department	Programs
City of Austin	Texas	2015	Austin Community Climate Plan	Office of Sustainability	Climate change, sustainability
City of Baltimore	Maryland	2013	Baltimore Climate Action Plan	Office of Sustainability	Climate action, sustainability
City of Boston	Massachusetts	2014	Greenovate Boston 2014 Climate Action Plan Update	Environment	Sustainable development, climate protection, environment
Broward County	Florida	2015	Climate Action Plan: Local Strategy to Address Climate Change	Environmental Planning and Community Resilience Division	Climate and energy program
City of Chicago	Illinois	2008	Climate Action Plan: Our City, Our Future	Department of Planning and Development/ Sustainable Development Division	Environment and sustainability
City of Dubuque	Iowa	2013	Dubuque Community Climate Action and Resiliency Plan 2013	Sustainable Dubuque	Climate action, energy
City of Eugene	Oregon	2010	A Community Climate and Energy Action Plan for Eugene	Office of Sustainability	Climate recovery
City of Los Angeles	California	2015	The pLAn and Los Angeles Climate Action Report: Updated 1990 Baseline and 2013 Emissions Inventory Summary	Office of Los Angeles Mayor Eric Garcetti	Sustainability, resiliency, and preparedness
Los Angeles County	Los Angeles	2015	Final Unincorporated Los Angeles County Community Climate Action Plan (CCAP) 2020	Department of Regional Planning	Planning, climate action
City of New Orleans	Louisiana	2017	Climate action for a Resilient New Orleans	Mayor's Office of Resilience and Sustainability	Resilience and sustainability
City of New York	New York	2017	1.5 °C: Aligning New York City with the Paris Climate Agreement	Mayor's Office of Sustainability	Climate and Energy
City of Philadelphia	PA	2016	Growing Stronger: Toward a Climate-Ready Philadelphia	Office of Sustainability	Energy benchmarking, climate adaptation planning
City of Portland	Oregon	2015	Climate Action Plan: Local Strategies to Address Climate Change (Portland and Multnomah County 2015 Climate Action Plan)	Portland Bureau of Planning and Sustainability	Planning, sustainability, climate action

(continued)

Table 6.1 (continued)

Local government	State	Year of release	Title of the plan	Department	Programs
City and County of San Francisco	California	2013	San Francisco Climate Action Strategy Department of Environment: 2013 Update	Department of Environment	Climate, energy, transportation, zero waste, urban forest and greening
Southeast Florida Regional Climate Change Compact Counties	Florida	2012	Southeast Florida Regional Climate Change Compact Counties - Regional Climate Action Plan	Southeast Florida Regional Climate Change Compact is a regional collaborative including Broward, Miami-Dade, Monroe, and Palm Beach Counties	Climate change

Resilience Spotlight: Michigan Cities Climate Preparedness and Planning

In contrasting Michigan cities approaches to climate resilience, there are apparent similarities and differences. Communities of Grand Haven, Grand Rapids, Ann Arbor, and Traverse City in Michigan among others have noteworthy climate resilience initiatives. To illustrate, City of Grand Rapids' sustainability plan encompasses all the aspects of sustainability and climate resilience planning, including climate adaptation and mitigation strategies (City of Grand Rapids 2016). To illustrate climate preparedness recommendations from a multi-stakeholder, long-term Grand Rapids' Climate Resiliency Report were incorporated into the city's 5-year sustainability plan (Alibašić 2017; City of Grand Rapids 2016).

On the other end of the spectrum, the City of Ann Arbor has a climate action plan and sustainability plan (City of Ann Abor 2012, 2013, 2015). Moreover, the City of Ann Arbor poured significant funding and staffing into sustainability and climate action planning (City of Ann Arbor 2015; Powers 2015; Stanton 2015). In Ann Arbor, the climate action plan includes greenhouse gas emissions inventory and climate action categories, setting goals to reduce community-wide greenhouse gas (GHG) emissions to 25% by 2025 and 90% by 2050 (City of Ann Arbor 2012, 2013). The Sustainability Action Plan includes direct climate mitigation and adaptation implications as it includes climate and energy outcomes and goals (City of Ann Arbor 2015). While approaches to climate resilience planning by these two cities appear divergent, the end goals and results are similar as both cities focus on climate change using both climate mitigation and adaptation strategies by embedding climate resilience in their plans. In the words of Matthew Naud, Environmental Manager for the City of Ann Arbor, "climate resilience and sustainability planning are inseparable, and our community values the city's attempting to address both climate adaptation and climate mitigation" (Email communication with author, January 25, 2018).

Traverse City developed a climate action plan with SEEDS, Inc., with funding from Department of Energy (City of Traverse City 2011). The program includes "ten strategies for action" and acts as:

> "a roadmap to meeting the climate and energy goals and objectives, while also being a living process that is adapted to changing needs," with "specific strategies for reducing GHGs, through fiscally sound methods, while also acting as a framework for decision making, thus acting as a guide to meeting established climate goals and objectives." (City of Traverse City 2011, p. 8)

Recently, the city leaders of Traverse City committed to a 100% renewable energy goal by 2020 (MCAN 2016).

In Grand Haven, local planners and administrators have developed a comprehensive and an impressive resilient master plan. The master plan among other issues covers climate change threats to the Great Lakes, using well-document and scientifically backed research to support the findings and recommendations. Some of the noted trends and risks are increased precipitation and storminess, a variability of lake water levels, and water temperature. As a coastal city, it is necessary to look at long-term trends and develop planning mechanism to deal with those obstacles in a comprehensive and dynamic, holistic approach. Embedding climate resilience strategies into the master plan is a practical, resilient planning approach. Moreover, in addition to featuring Grand Haven, resilient Michigan organization lists other communities in Michigan with resilient master plans in place, including cities of Holland, Beaver Island, Bridgman, East Jordan, Ludington, Macomb/St. Clair, and St. Joseph (Resilient Michigan n.d.). To illustrate, Resilient Monroe Resource Atlas is a land-use master planning and community design project for the City of Monroe, Frenchtown Charter Township, and Monroe Charter Township. It includes, among other tools and analyses, a review of the urban heat island effect, heat sensitivity and exposure assessment, flood vulnerability assessment, and drainage stormwater management (Resilient Monroe 2013). Authors of the report noted that:

> climate scientists say that the Monroe Community and southeast Michigan can expect more frequent storms of increasing severity in the decades ahead. The total amount of rainfall is also likely to increase. However, climate models suggest that the precipitation will be more concentrated in the winter, spring and fall seasons as well as localized intense storms at almost any time. (Resilient Monroe 2013, pp. 8–10)

In the recent years, both the planning and investments in infrastructure made by cities were tested with the flood events and the extreme heat waves in the summers. Communities have made preemptive investments, in sustainability and emergency planning to avoid further costly damages to the infrastructure and resources and implementing a variety of climate mitigation and adaptation strategies, including but not limited to:

- Developing energy conservation and efficiency strategies to reduce energy consumption and demand throughout the organization.
- Moving the power demand to purchase and production from renewable sources, as an attempt to diversify energy sources, as an essential step toward local resilience and for greenhouse gas reduction.
- Reducing total fuel consumption.
- Setting a goal to increase its tree canopy cover and diversify the type of tree species planted.

Local governments have been partnering with local nonprofits, grass root organizations, residents, and academic institutions and developing climate change assessments, data reports, and resilience plans to further concentrate on climate, energy issues, economy, transportation, and infrastructure and to inform decision-makers in the areas of sustainability, ordinances, policies, and adaptation and mitigation strategies. The resilience plans acknowledge the facts of climate change and serve to prepare the community and make it more agile and adaptive to extreme events and disasters.

Climate Resilience and Economic Development

There are potential positive outcomes from linking resilience planning, sustainability, and organizational efficiency to promote resilience and policies to decrease carbon emissions and lower costs and improve economic development and growth opportunities. Fitzgerald (2010) suggested the "cities can employ economic development strategies to support the development of renewable energy and clean carbon-reducing technologies" and moreover note the linkages between the sustainability and climate change initiatives (pp. 8, 178). Importantly, it is incumbent for local governments to recognize the realities of climate change in resilience reports and plans. As a way of illustration, in its resilience plan, the Resilient Macomb (2016) included the following statement "climatologists are projecting that extreme weather events will increase in frequency and intensity in Southeast Michigan" (p. 2). Linking those climate change concerns, supported with data by the realities of impact on jobs, business, and economic development, is essential to resilience planning. As noted in the Resilient Macomb (2016) plan, "Lake St. Clair enables about 660,000 jobs in manufacturing, farming, mining, and energy production to exist. Tourism is responsible for about 57,000 additional jobs" (p. 3).

Embracing New Technologies and Systems

The importance of adopting new technology and system improvement is exemplified through local governments' ability to continue providing services in time of crises. The local units of government regularly evaluate the technological advancements and are willing to embrace new technologies. However, it was clear that willingness to serve as a beta site for new technologies diminished with the potential that the city may incur risk as a result of testing new systems or technologies. Fiksel (2006) argued that the assessment of "interactions among interdependent systems requires new tools to capture the emergent behaviors and dynamic relationships that characterize complex, adaptive systems" (p. 17). A greater level of collaboration in the technological systems deliver synergies and improvement to organizations, from sharing of knowledge, training opportunities, engineering practices, and cost-savings from measures to increase resilience.

Building Resilience Through Transformation and Awareness

There are compelling reasons local governments to embrace and benefit from climate resilience planning from sustainable energy outcomes, change in culture, and a transformation in operations and service delivery. There is an overall belief that sustainable energy management has a positive impact on operations and meets community expectations for change. Policy makers and administrators embrace ambiguities in adopting transformational changes and strategies in organizations. Transformative measures are undertaken by organizations to build a stronger and more resilient community ready to respond to changing demands and surrounding economic and environmental threats and uncertainty. The transformation toward sustainability is viewed as an opportunity for building more resilient organizations and communities.

Resilient County Spotlight: Broward County Embedding Climate Resilience

A vital component of a successful implementation of sustainability and resilience planning is the organizational commitment. In the organizational chart of the Broward County, Climate Change Program is featured under the Environmental Planning and Community Resilience Division (Broward County 2017). Among many different initiatives, the division overseas climate change, Go Green, energy and sustainability programs (Broward County 2015). Dr. Jennifer Jurado, Director of the Environmental Planning and Community Resilience Division, finds the Broward County's "staffing and financial commitment to climate resilience and sustainability planning a critical element to a more resilient county and the region" (Email to the author, January 29, 2018).

Summary

The focus of this chapter was the interconnection between sustainability and resilience planning and strategies in local governments. If properly utilized, tracked, measured, and compared to actual budgetary results and fiscal performance, resilience and sustainability strategies produce a tangible, long-term effects on the overall effectiveness of service delivery. The ultimate goals of resilience planning are an improvement of the governance, reduction of the cost of operations, a decrease of the environmental impact, and positive social effects on communities. Analyzing the overall effectiveness of climate preparedness and its implications on social, economic, ecological, and governance of organizations leads to better understanding of and confidence in local governments' resilience planning.

Moving resilience planning to a regional level allows for outcome-driven partnership and sharing of responsibilities and resources. By using a dynamic approach to resilience planning, local governments steadily adapt to shifting economic, environmental, social, and governing conditions. Resilient organizations and communities continually build upon existing plans, layering and preparing to adapt and mitigate. Accordingly, public service practitioners examine the current policies in place to identify strategies and targets to meet climate resilience outcomes. Embedding climate resilience strategies into existing plans is an efficient way of committing to the climate resilience action, as long as local governments tie the implementation and projects to their budget process.

The climate preparedness, readiness, and resilience planning have come to the forefront for local government to create a more resilient future for communities. Resilience is observed and implemented contextually as an instrument leading to an improved governance of environmental, social, and economic resources. Moreover, the commitment to sustainability and resilience will predictably lead to improved social and environmental outcomes. Worrell and Appleby (2000) defined stewardship as the concept of "responsible use of natural resources" and acceptance of "significant answerability to society" (p. 275). There is a balance, and an argument may be made that similar to the private sector, sustainability and resilience for municipalities begin with the economic bottom line. As the layers of the organization embrace sustainability and resilience strategies and polices, embedding them within structures, good governance, along with culture change and transformation lead to improvements and micro- and macro-level solutions to climate-related issues and problems. The local governments in their capacity play an crucial role in the policy development, program implementation and the practical applications of climate resilience plans and strategies.

Further Discussions

- Define and examine climate resilience, climate preparedness, climate mitigation and adaptation, and sustainable and resilient communities.
- Analyze communities' plans to reduce greenhouse gas emissions and to address the climate change threats.
- Discuss the impact of climate change on the nation, state, and region.
- Assess the necessary elements for climate preparedness and resilience planning in communities.

References

Alibašić H (2014) Planning and implementing climate resiliency in Grand Rapids. The Review, May–June. Michigan Municipal League. Retrieved from http://www.mml.org /resources/publications/mmr/issue/may-june2014/review-mayjune2014-online.pdf

Alibašić H (2017) Measuring the sustainability impact in local governments using the Quadruple Bottom Line. Int J Sustain Policy Prac 13(3):37–45

Alibašić H (2018a) Leading climate change at the local government level. In: Farazmand A (ed) Global encyclopedia of public administration, public policy, and governance. Springer International AG, Zurich. https://doi.org/10.1007/978-3-319-31816-5_3428-1

Alibašić H (2018b) Ethics of resiliency in crisis management. In: Farazmand A (ed) Global encyclopedia of public administration, public policy, and governance. Springer International AG, Zurich. https://doi.org/10.1007/978-3-319-31816-5_3426-1

Baltimore Office of Sustainability (2013) Baltimore climate action plan. Retrieved from http://www.baltimoresustainability.org/wp-content/uploads/2015/12/BaltimoreClimateActionPlan.pdf

Baucum J (2017) Pensacola climate change task force meets to curb local global warming impact. Pensacola News Journal. Retrieved from: http://www.pnj.com/

story/news/politics/2017/06/02/pensacola-climate-change-task-force-trump-paris-accord/362602001/
Broward County (2015) Climate action plan: local strategy to address climate change. Retrieved from http://www.broward.org/NaturalResources/ClimateChange/Documents/BrowardCAPReport2015.pdf
Broward County (2017) Environmental planning and community resilience division. Organizational chart. Retrieved from http://www.broward.org/NaturalResources/Documents/orgchartepcrd.pdf
Brugmann J (2012) Financing the resilient city. Environ Urban 24:215–232. https://doi.org/10.1177/0956247812437130
Burch S (2010) In pursuit of resilient, low carbon communities: an examination of barriers to action in three Canadian cities. Energy Policy 38:7575–7585. https://doi.org/10.1016/j.enpol.2009.06.070
Byrne J, Hughes K, Toly N, Wang Y (2006) Can cities sustain life in the greenhouse? Bull Sci Technol Soc 26:84–95. https://doi.org/10.1177/0270467606287532
City of Ann Arbor (2012) City of Ann Arbor climate action plan. Retrieved from: http://www.a2gov.org/departments/systems-planning/planning-areas/energy/Documents/CityofAnnArborClimateActionPlan_low%20res_12_17_12.pdf
City of Ann Arbor (2013) Sustainability framework. Retrieved from http://www.a2gov.org/departments/systems-planning/planning-areas/climate-sustainability/sustainability/Pages/SustainabilityFramework.aspx
City of Ann Arbor (2015) Sustainability action plan draft. Retrieved from http://www.a2gov.org/departments/systems-planning/planning-areas/climate-sustainability/Sustainability-Action-Plan/Documents/SAP%20-%20DRAFT%20-%20%20July%202015%20-%20web.pdf
City of Austin (2014) Report on Climate Resilience Resolution 20131121-060; From Lucia Athens, Chief Sustainability Officer; Zach Baumer, Climate Program Manager. Retrieved from http://www.austintexas.gov/sites/default/files/files/Toward_a_Climate_Resilient_Austin.pdf
City of Austin (2015) Austin community climate plan. Retrieved from http://www.austintexas.gov/sites/default/files/files/Sustainability/FINAL_-_OOS_AustinClimatePlan_061015.pdf
City of Boston (2014) Greenovate Boston: 2014 climate action plan update. Retrieved from https://www.boston.gov/sites/default/files/greenovate_boston_2014_cap_update.pdf
City of Chicago (2008) Climate action plan: our city, our future. Retrieved from: http://www.chicagoclimateaction.org/filebin/pdf/finalreport/CCAPREPORTFINALv2.pdf
City of Dubuque (2013) Dubuque community climate action & resiliency plan 2013. Sustainable Dubuque. Retrieved from http://cityofdubuque.org/DocumentCenter/View/18359
City of Dubuque (2017) Sustainable Dubuque. Retrieved from: http://www.sustainabledubuque.org/
City of Eugene (n.d.) Climate recovery summary. Retrieved from: https://www.eugene-or.gov/3210/Climate-Recovery-Summary
City of Eugene (2010) A community climate and energy action plan for Eugene. Retrieved from: https://www.eugene-or.gov/Archive/ViewFile/Item/80
City of Fayetteville (n.d.) Arkansans can take steps to respond to climate change. Retrieved from http://www.fayetteville-ar.gov/DocumentCenter/View/14890
City of Grand Rapids (2015) Fifth Year Sustainability Plan Progress Report. Alibašić H and Gosztyla D (eds.). Retrieved from: https://www.grandrapidsmi.gov/files/assets/public /departments/office-of-sustainability/reports-and-documents/office-of-sustainability/5th-year-progress-report-sustainability-plan.pdf
City of Grand Rapids (2016) Sustainability Plan FY2017-FY2021. Office of Energy and Sustainability Alibašić H (ed.). Retrieved from https://www.grandrapidsmi.gov/files/assets/public/departments/office-of-sustainability/reports-and-documents/office-of-sustainability/sustainability-plan-fy17-fy21.pdf
City of Los Angeles (2015a) The pLAn. Office of Mayor Eric Garcetti. Retrieved from http://plan.lamayor.org/wp-content/uploads/2017/03/the-plan.pdf
City of Los Angeles (2015b) Los Angeles climate action report: updated 1990 baseline and 2013 emissions inventory summary. https://www.lamayor.org/sites/g/files/wph446/f/landing_pages/files/pLAn%20Climate%20Action-final-highres.pdf
City of New York (2017) 1.5°C: aligning New York City with the Paris Climate Agreement. Mayor's Office of Sustainability. Retrieved from https://www1.nyc.gov/assets/sustainability/downloads/pdf/1point5-aligning-nyc-with-paris-agreement.pdf
City of New Orleans (2017) Climate action for a Resilient New Orleans. Mayor's Office of Resilience and Sustainability. Retrieved from https://www.nola.gov/nola/media/Climate-Action/Climate-Action-for-a-Resilient-New-Orleans.pdf
City of Traverse City (2011) City of Traverse City climate action plan. by Solomon Townsend, Barton, Kirk and Michael Powers. SEEDS, Inc. Traverse City.. Retrieved from http://www.traversecitymi.gov/downloads/climateactionplanfeb2011.pdf
City of Philadelphia (2016) Growing stronger: toward a climate-ready Philadelphia. Retrieved from https://beta.phila.gov/media/20160504162056/Growing-Stronger-Toward-a-Climate-Ready-Philadelphia.pdf
City of Philadelphia (2017) Publications. Retrieved from https://beta.phila.gov/departments/office-of-sustainability/publications/
City of Portland (2015) Climate action plan: local strategies to address climate change (Portland and Multnomah County 2015 Climate Action Plan). Retrieved from https://www.portlandoregon.gov/bps/article/531984

City of Portland and Mulnomanh County (2017) Climate Action Plan Progress Report. Retrieved from https://www.portlandoregon.gov/bps/article/636700

City and County of San Francisco (2013) San Francisco climate action strategy. Department of Environment. Retrieved from https://sfenvironment.org/sites/default/files/engagement_files/sfe_cc_ClimateActionStrategyUpdate2013.pdf

City and County of San Francisco (2017) Climate plans and reports. San Francisco Environment. Retrieved from: https://sfenvironment.org/climate-plans-reports

Climate Mayors (2017) Cities adopt the Paris Climate Agreement goals. Retrieved from http://climatemayors.org/

County of Los Angeles (2015) Final Unincorporated Los Angeles County Community Climate Action Plan (CCAP) 2020. Department of Regional Planning. Retrieved from http://planning.lacounty.gov/assets/upl/project/ccap_final-august2015.pdf

Davenport C (2017) Counseled by industry, not staff, E.P.A. chief is off to a blazing start. The New York Times. Retrieved from https://www.nytimes/com/2017/07/01/us/politics/trump-epa-chief-pruittregulations-climate-change.html

Dong W, Liu Z, Liao H, Tang Q, Li X (2014) New climate and socioeconomic scenarios for assessing global human health challenges due to heat risk. Clim Chang 130:505–518

Fiksel J (2003) Designing resilient, sustainable systems. Environ Sci Technol 37:5330–5339. https://doi.org/10.1021/es0344819

Fiksel J (2006) Sustainability and resilience: toward a systems approach. Sustain Sci Prac Policy 2(2):14–21. Retrieved from http://ejournal.nbii.org

Fitzgerald J (2010) Emerald cities: urban sustainability and economic development. Oxford University Press, New York

Fitzgerald BG, O'Doherty T, Moles R, O'Regan B (2012) A quantitative method for the evaluation of policies to enhance urban sustainability. Ecol Indic 18:371–378. https://doi.org/10.1016/j.ecolind.2011.12.002

Fletcher C (2013) Climate change: what the science tells us. Wiley, Heboken

Gordon DJ (2016) The politics of accountability in networked urban climate governance. Glob Environ Polit 16(2):82–100

Hallegatte S, Henriet F, Corfee-Morlot J (2011) The economics of climate change impacts and policy benefits at city scale: a conceptual framework. Clim Chang 104:51–87. https://doi.org/10.1007/s10584-010-9976-5

Intergovernmental Panel on Climate Change (IPCC) (2014) Summary for policymakers. In: Field Christopher B, Barros VR, Dokken DJ, Mach KJ, Mastrandrea MD, Bilir TE, Chatterjee M, Ebi KL, Estrada YO, Genova RC, Girma B, Kissel ES, Levy AN, MacCracken S, Mastrandrea PR, White LL (eds) Climate change 2014: impacts, adaptation, and vulnerability. Part A: global and Sectoral aspects. Contribution of working group II to the fifth assessment report of the intergovernmental panel on climate change. Cambridge University Press, Cambridge, pp 1–32

Karl TR, Melillo JM, Peterson TC (eds) (2009) Global climate change impacts in the United States. US Global Change Research Program. Retrieved from https://downloads.globalchange.gov/usimpacts/pdfs/climate-impacts-report.pdf

Kling GW, Hayhoe K, Johnson LB, Magnuson JJ, Polasky S, Robinson SK, Shuter BJ, Wander MM, Wuebbles DJ, Zak DR, Lindroth RL, Moser SC, and Wilson ML (2003) Confronting climate change in the Great Lakes Region: impacts on our communities and ecosystems. Union of Concerned Scientists, Cambridge, Massachusetts, and Ecological Society of America, Washington, D.C. Retrieved from: https://mybb.gvsu.edu/bbcswebdav/pid-3951258-dt-content-rid-36059702_1/courses/GVPA671.01.201710/GVPA671.01.201330_ImportedContent_20130423031707/greatlakes_final.pdf

Lindfield M (2010) Cities: a global threat and a missed opportunity for climate change. Environ Urban ASIA 1(2):105–129. https://doi.org/10.1177/097542531000100202

Mach KJ, Mastrandrea MD, Bilir E, Field CB (2016) Understanding and responding to danger from climate change: the role of key risks in the IPCC AR5. Clim Chang 136:427–444. https://doi.org/10.1007/s10584-016-1645-x

Malcolm JR, Liu C, Neilson RP, Hansen L, Hannah L (2006) Global warming and extinctions of endemic species from biodiversity hotspots. Conserv Biol 20:538–548

Michigan Climate Action Network (MCAN) (2016) Traverse City commits to 100% clean energy by 2020. Retrieved from http://www.miclimateaction.org/traverse_city_commits_to_100_percent_clean_energy_by_2020

Pecl G, Araújo M, Bell J, Blanchard J, Bonebrake T, Chen I-C et al (2017) Biodiversity redistribution under climate change: impacts on ecosystems and human well-being. Science 355(6332). https://doi.org/10.1126/science.aai9214

Powers SD (2015) City of Ann Arbor Administrator General Fund Budget Memo. Retrieved from http://media.mlive.com/annarbornews_impact/other/city_admin_FY16_budget.pdf

Resilient Macomb (2016) Planning for coastal resilience in Macomb County, Michigan. Retrieved from http://www.resilientmichigan.org/downloads/macomb_coastalresiliencyreport_web.pdf

Resilient Michigan (n.d.) Resilient Michigan Communities. Retrieved from; http://www.resilientmichigan.org/communities.asp

Resilient Monroe (2013) Resilient Monroe Resource Atlas. Retrieved from http://www.resilientmichigan.org/downloads/resilient_monroe_resource_atlas_20131015.pdf

Silva RA, West JJ, Zhang Y, Anenberg SC, Lamarque JF, Shindell DT, Collins WJ et al (2013) Global premature mortality due to anthropogenic outdoor air pollution and the contribution of past climate change. Environ Res Lett 8(034005):1–11. https://doi.org/10.1088/1748-9326/8/3/034005

Segan DB, Hole DG, Donatti CI, Zganjar C, Martin S, Buthart SHM, Watson JEM (2015) Considering the impact of climate change on human communities significantly alters the outcome of species and site-based vulnerability assessments. Divers Distrib 21:1101–1111

Southeast Florida Regional Climate Change Compact Counties (2012) Regional climate action plan. A region responds to a changing climate. Retrieved from https://southeastfloridaclimatecompact.files.wordpress.com/2014/05/regional-climate-action-plan-final-ada-compliant.pdf

Stott PA, Christidis N, Otto FEL, Sun Y, Vanderlinden J, van Oldenborgh GK et al (2016) Attribution of extreme weather and climate-related events. Wiley Interdiscip Rev Clim Chang 7(1):23–41. https://doi.org/10.1002/wcc.380

Stanton R (2015) See details of Ann Arbor's proposed $380M city budget. Mlive. Retrieved from http://www.mlive.com/news/ann-arbor/index.ssf/2015/04/ann_arbors_proposed_380m_city.html

Tesla N (1900) The problem of increasing human energy: with special reference to the harnessing of the sun's energy. The Century Magazine. Retrieved from http://www.teslauniverse.com/nikola-tesla-article-the-problem-of-increasing- human-energy

The United States Conference of Mayors (USCM) (2005) Mayors climate protection agreement. Retrieved from https://www.usmayors.org/mayors-climate-protection-center/

The United States Conference of Mayors (USCM) (2017) 85th Annual Meeting 2017 adopted resolutions. Retrieved from http://legacy.usmayors.org/resolutions/85th_Conference/proposedcommittee.asp?committee=Energy

The United States Environmental Protection Agency (EPA) (2018) Green power partnership. Green power partnership top partner rankings. Retrieved from: https://www.epa.gov/greenpower/green-power-partnership-top-partner-rankings

The State, Local, and Tribal Leaders Task Force on Climate Preparedness and Resilience (Task Force) (2014) Recommendations to the President. Retrieved from https://obamawhitehouse.archives.gov/sites/default/files/docs/task_force_report_0.pdf

The World Bank (2012) Turn down the heat: why a 4°C warmer world must be avoided. A report for the World Bank by the Potsdam Institute for Climate Impact Research and Climate Analytics. Retrieved from https://mybb.gvsu.edu/bbcswebdav/pid-3951253-dt-content-rid-36059696_1/courses/GVPA671.01.201710/GVPA671.01.201330_ImportedContent_20130423031707/Turn_Down_the_heat_Why_a_4_degree_centrigrade_warmer_world_must_be_avoided.pdf

Urban MC (2015) Accelerating extinction risk from climate change. Science 348:571–573

U.S. Department of State (2017) Communication regarding intent to withdraw from Paris Agreement. Office of the Spokesperson. Retrieved from: https://www.state.gov/r/pa/prs/ps/2017/08/273050.htm

U.S. Global Change Research Program (USGCRP) (2017) In: Wuebbles DJ, Fahey DW, Hibbard KA, Dokken DJ, Stewart BC, Maycock TK (eds) Climate science special report: fourth National Climate Assessment, volume I. U.S. Global Change Research Program, Washington, DC, p 470. https://doi.org/10.7930/J0J964J6

Visser H, Petersen AC, Ligtvoet W (2014) On the relation between weather-related disaster impacts, vulnerability and climate change. Clim Chang 125:461–477

West Michigan Environmental Action Council (WMEAC) (2013) Grand Rapids Climate Resiliency Report. Retrieved on September 11, 2017, from: https://wmeac.org/wp-content/uploads/2014/10/grand-rapids-climate-resiliency-report-master-web.pdf

Worrell R, Appleby MC (2000) Stewardship of natural resources: definition, ethical and practical aspects. J Agric Environ Ethics 12(3):263–277. Retrieved from http://search.proquest.com/docview/196564943

Evaluating Tools and Resources for Sustainability and Resilience Planning

7

"Furthermore, towns and cities with their monuments, vast constructions, and large buildings, are set up for the masses and not for the few. Therefore, united effort and much co-operation are needed for them." ***Ibn Khaldun, A. A. bin M (1406, 1969). The Muqaddimah: An introduction to history; In three volumes.***

Key Questions

The goal of the seventh chapter of the book is to answer the following underlying assumptions and questions:

- What sustainability and climate preparedness resources do local governments need to plan for sustainable communities and to achieve a maximum organizational resilience?
- What tools do communities require for climate preparedness and resilience?
- What tools and resources are available for sustainability and resilience planning?
- What carbon inventory and forecasting resources are recommended for the organizations?
- What templates or outlines are appropriate for local governments' sustainability and resilience planning?
- How do communities integrate climate readiness into emergency preparedness?

Introduction

The seventh chapter of the book provides a cursory and illustrative review of available resources, tools, and assistance for sustainability and resilience planning for local governments to attain maximum resilience and sustain communities. Beyond consulting services, resources and tools are available at no or very low cost and require internal coordination, staffing, and management. The chapter provides a brief overview of accessible national, state-level, and regional resources and organizations supporting the climate preparedness and resilience efforts, including available tools for a carbon footprint inventory or greenhouse gas emission reporting. The list is based on the past professional experience and publicly available information. It is intended as a possible first step for organizations interested in starting the sustainability and resilience planning. It is neither comprehensive nor final, as there are numerous organizations, resources, and tools available to local governments in each state, nationally, and internationally. An illustrative overview of a template of a sustainability and resilience plan using the Quadruple Bottom Line approach is provided. A template for integrating a climate readiness into emergency preparedness and disaster mitigation plan is offered.

Sustainability and Resilience Networks and Organizations

A brief overview of several national, regional, and statewide resources and organizations available to support sustainability and resilience

H. Alibašić, *Sustainability and Resilience Planning for Local Governments*, Sustainable Development Goals Series, https://doi.org/10.1007/978-3-319-72568-0_7

planning is provided. Some federal and state agencies, national and international organizations afford access to planning resources for free or for members. For instance, the US Environmental Protection Agency Pollution Prevention (P2) program offers beneficial links and resources.

Federal Agencies' Resilience Resources

The US Environmental Protection Agency GHG calculator One of the most relevant aspects of the local governments' commitment to reducing the harmful environmental impact is their ability to reduce the GHG emissions in operations. The resources and tools provided by the US Environmental Protection Agency are particularly beneficial for organizations lacking resources to retain consultants and relying on internal staff for measuring and reporting carbon footprint. The website includes case studies and tools for pollution prevention (EPA 2017). The P2 GHG emission calculator quantifies and inventories annual greenhouse gas (GHG) emissions and reductions from projects aimed at reducing pollution and enhancing resilience. The excel sheet tool is divided into tabs, with each using formula to calculate and aggregate the CO2 emissions, cost savings of GHG emissions, electricity management, green energy, solvent manufacturing, stationary sources, greening chemistry, mobile sources, and water management (EPA 2017). The P2 cost calculator's unique feature includes a direct conversion of the GHG reductions to the related cost savings, drawing from data in the P2 cost-saving calculator (EPA 2017). In 2017, the climate change information and resources were removed from the EPA website in 2017. However, cities in the United States are preserving decades of the EPA research on climate change on their websites (City of Houston 2017; City of Milwaukee 2017).

The National Oceanic and Atmospheric Administration (NOAA) offers a US climate resilience tool as a one-stop place for resilience planning with case studies, comprising of the climate explorer, funding opportunities for resilience, and topics on built environment, coasts, food, marine, energy, health, transportation, water, and tribal nations (NOAA n.d.). While some of the tools are not strictly developed by NOAA and are hosted on other organizations' sites, the NOAA's website features 75 tools. Each tool serves as a guide to assist the organizations and communities with the various aspects of climate preparedness. The tools vary from Coastal Flood Exposure Mapper; Quick Report Tool for Socioeconomic Data, with the access to economic and demographic data for multiple coastal jurisdictions; and many other valuable resources (NOAA 2017). The NOAA's Great Lakes Environmental Research Laboratory observes, monitors, and forecasts accurate and up-to-date Great Lakes water levels (NOAA 2018).

(Inter)national Networks Supporting Local Sustainability and Climate Resilience Initiatives

The notion of climate resilience and preparedness at a local level of government began to take shape in recent years and was advanced by the President Obama's State, Local, and Tribal Leaders Task Force on Climate Preparedness and Resilience (Task Force 2014). Moreover, with the 100 Resilient Cities grants, 100 cities around the globe were able to hire the Chief Resilience Officers (CROs) to start plans for climate resilience (100 Resilient Cities 2018). The 100 Chief Resilience Officers is an initiative aimed at building infrastructure in cities to plan for climate change, through climate resilience policies and programs. By enabling local governments to employ a full staff person dedicated to climate resilience, more cohesive efforts are being deployed at subnational levels, allowing a community-based initiative to become self-sustained and embedded.

The US Conference of Mayors through Mayors' Climate Protection Center and a myriad of similar sustainability initiatives encourages member cities and mayors to take a leadership role in addressing the threats of climate change and to meeting carbon reduction goals (USCM 2018). Comparably, the National League of Cities engages its members to take proactive actions on environmental, sustainability, and climate change issues and affords its members with tools and resources for climate engagement activities (NLC 2017, 2018). The Great Lakes and St. Lawrence Cities Initiative (GLSLCI) enables networking, sharing of case studies and best practice, and advocacy opportunities for its members on an array of issues related to water resources' protection, including climate change planning, sustainability, and energy (GLSLCI 2016). Additionally, a nonprofit organization, the Architecture 2030 (2017a, b), was established in 2002, with the "mission to rapidly transform the global built environment from the major contributor of greenhouse gas (GHG) emissions to a central part of the solution to the climate crisis" (para. 1). It oversees the 2030 Districts, the private-public challenge for the reduction of energy use and carbon emissions in fifteen cities in the United States and Canada: Seattle, Cleveland, Los Angeles, Pittsburgh, Denver, San Francisco, Stamford, Dallas, Toronto, Albuquerque, San Antonio, Grand Rapids, Austin, Portland ME, and Ithaca. The commercial building owners with over 290 million square feet of commercial buildings in 2030 District designated cities commit to reducing greenhouse gas emissions, water conservation, and energy consumption (Architecture 2030 2015, 2017a, b).

Fiscal constraints and the uncertainties of a global economy impact the local governance and the ability of cities to deliver outcomes. viewed as significant disruptions. Local governments engaged in climate preparedness and resilience efforts adapt, transform, and accept significant disruptions as they continue to provide services without interruptions. In a crisis, community resilience related initiatives and planning become an opportunity for local governments in changing budgetary priorities, emergency management, operations, and processes. A collection of national organizations including the Local Governments for Sustainability (ICLEI), World Wildlife Fund (WWF), US Green Building Council (USGBC), and National League of Cities (NLC) formed the Resilient Communities for America (RC4A) (RC4A n.d.). The RC4A group recognized that local governments are on the front lines of the climate change challenges and must respond to climate change pressures (RC4A n.d.). Another engaging network for local governments, which supports the strategies for reduction of carbon footprint is the Compact of Mayors, with a set of targets set for cities participating in the agreement (Compact of Mayors 2018; Global Covenant of Mayors n.d.). With the US administration's decision to withdraw from Paris climate agreement, the local governments remain committed to climate change preparedness and resilience goals (Alibašić, 2018a, c; Compact of Mayors 2018; Sengupta et al. 2017; The Telegraph Reporters 2017). Additionally, within their capacity, the corporations are committing resources to prepare and ready their operations for the climate change-related risks (Alibašić 2018b).

Resilient Organizations and Initiatives Spotlight

Local Government for Sustainability (ICLEI) With global membership, ICLEI contains one of the most extensive networks of local governments. Since 1991 ICLEI has been offering planning services on sustainability and climate change to local government leaders and administrators (ICLEI 2018a). In addition to providing the greenhouse gas protocols, ICLEI offers tools for its local government members, to assist them in preparing for climate change impacts and to manage emissions (ICLEI 2018b).

Urban Sustainability Directors Network (USDN) The USDN is a peer-to-peer network where members share information with each other and share best practices on sustainability planning (USDN 2018).

100 Resilient Cities (100RC) The 100 Resilient Cities 100RC) was created and funded by the Rockefeller Foundation. Only cities accepted into the network are afforded with the financial and logistical support to create a position of a "chief resilience officer" and expertise to develop a "robust resilience strategy," with an access to resources for implementation of resilience strategies and information sharing between member cities (100RC 2018).

The Nature Conservancy The organization features links to the climate adaptation and planning tools, including the Great Lakes Water Levels Dashboard, Climate Wizard with visual aid representing temperature trends, Great Lakes Coastal Resilience Planning Guide, and Digital Coast from NOAA Coastal Services Center with geospatial data information (The Nature Conservancy n.d.-a, b, c).

Resilient Michigan *Resilient Michigan* is a collaborative effort "developed by the Land Information Access Association (LIAA), a nonprofit community service and planning organization headquartered in Traverse City, Michigan" (Resilient Michigan n.d.). According to the available website information, the Planning for Resilient Communities "fosters and supports community-wide planning efforts that promote community resilience in the face of rapid economic changes and increasing climate variability" (Resilient Michigan n.d.). It provides a practical handbook and a resource guide within the following nine resilience themes: local governance and leadership, gray and green infrastructure, transportation, local food and food systems, housing and neighborhoods, natural resources, public health, coastal processes, and energy (Resilient Michigan 2017).

Regional and Multistate Resources

Northern Gulf of Mexico Sentinel Site Cooperative The Gulf TREE (Tools for Resilience Exploration Engine) assists the local government planners and administrators, and natural resource managers with climate preparedness and readiness planning. The search engine is designed as a decision-support tree framework assisting users in the Gulf of Mexico region to identify the appropriate climate resilience tool, used to evaluate and analyze the science-based data to prepare for hazards and resilience for the coastal areas. The Gulf TREE climate change and resilience exploration engine is a collaborative endeavor between the Northern Gulf of Mexico Sentinel Site Cooperative (SSC), the Gulf of Mexico Climate and Resilience Community of Practice (CoP), and the Gulf of Mexico Alliance (GOMA) team (Northern Gulf of Mexico Sentinel Site Cooperative 2018).

Sustainability and Resilience Plan Template

At a minimum, local governments' sustainability and resilience plan should consist of the following elements:

- Setting specific goals, generally and categorically, defines the desired outcome related to sustainability and resilience.
- Aligning goals with organizational objectives.
- Establishing the timetable for accomplishing resilience and sustainability-related objectives.
- Identifying internal and external stakeholders.
- Prioritizing projects with most impacts, as a case in point: waste minimization, recycling, public safety, transportation, energy, energy efficiency and renewables, water resource protection, and housing.

- Proposing and incorporating targets and outcome champions.
- Establishing greenhouse gas emissions, carbon footprint, and climate resilience goals, objectives, and targets.
- Determining measurements and Quadruple Bottom Line impacts.
- Creating a table with specific economic, social, governance, and environmental outcomes and goals and corresponding metrics.
- Defining implementation strategy and management structure/connection to the budget process.
- Mapping out the internal and external stakeholders and public outreach and campaign plan.

The proposed plan may be set in the following order: *themes >> goals >> outcome >>> targets*

The Quadruple Bottom Line (QBL) provides the overarching pillars: **economic**, **environmental**, **social**, and **governance**. The specific themes of this plan are then housed under one of the four QBL pillars, and specific targets are categorized under separate goals and outcomes. While understandably the four components are associated with sustainability, they are used for developing a sustainability and ultimately a resilience plan. The four QBL pillars provide an outline for identifying four categories of sustainability and resilience initiatives regardless of what the community decided to call the plan (strategic sustainability plan, climate adaptation and mitigation plan, climate action plan or ideally, resilience plan).

Under goals, there are multiple options and opportunities, tailored to specific needs of the community.
A sample of suggested goals may include the following.

1. Economic Growth and Opportunities
2. Resilient Environment
3. Resilient and Safe Neighborhoods
4. Resilient Assets: Infrastructure, Buildings, Utilities, Balanced Transportation
5. Good Governance: Accountability, Transparency, Accessibility, Community Input, and Fiscal Resilience

Each outcome provides a broader spectrum of objectives to be correlated with targets, and each is under a specific goal and theme. The plan may be color-coded for ease in identifying specific objectives; in the sample provided in this chapter, blue represents the economic pillar, green is for environmental, purple is for social, and red color characterizes the governance theme.

An illustrative template of a sustainability and resilience plan with targets using the Quadruple Bottom Line approach is provided here.

Economic Theme 1: Economic Growth and Opportunity

Goal 1: Encourage sustainable economic growth and create resilient economy to facilitate job creation.

Outcome 1.1:

Create resilient, entrepreneur-focused economic development strategies leveraging the local government resources and building the internal and external infrastructure required to support job creation and maintain the economic vitality of the community and the region.

Targets:

1. Support the [DOLLAR AMOUNT] in private investment by [YEAR, DATE].
2. Ensure that [PERCENTAGE] of jobs created or retained with incentives will be permanent, full-time employment with benefits annually.

Environmental Theme 1: Resilient Environment

Goal 1: Reduce energy and carbon footprint and strengthen organizational and community-wide climate resilience

Outcome 1.1:

Implement initiatives to counteract the effects of carbon emissions to provide for a cleaner, greener, more resilient, and climate-ready community.

Targets:

1. Achieve [PERCENTAGE] of energy production from renewable energy sources (wind, solar, or other non-traditional energy sources) by [YEAR, DATE].
2. Implement [PERCENTAGE] of on-site stormwater management to all new infrastructure projects by [YEAR, DATE].
3. Reduce the city's and community's greenhouse gas (GHG) emissions to [PERCENTAGE] below [YEAR] levels by [YEAR].
4. Double the water reuse and recovery by [YEAR, DATE] from [NUMBER] of gallons/day to [NUMBER] of gallons/day.

Social Theme 1: Resilient and Safe Neighborhoods

Goal 1: Neighborhood partnership and collaboration

Outcome 1.1: Collaborate with cross-sector groups to promote safety and reduce the occurrence of crime.

Targets:

1. Reduce the number of offenses below the [NUMBER] annually.

Goal 2: Safety operations and programs

Outcome 1.2.: Implement cost-effective, data-driven programs designed for high-risk groups and environments to promote safety, prepare for emergencies, and install and maintain city equipment and systems that ensure a safe and resilient environment for residents and businesses.

Targets:

1. Remove graffiti in the city within [TIMEFRAME (DAYS/HOURS)] from the time of notification [PERCENTAGE] of the time.

Governance Theme 1: Resilient Assets- Infrastructure, Buildings, Utilities, and Balanced Transportation

Goal 1: Maintain sustainable asset management

Outcome 1.1: Implement an integrated, lifecycle investment approach to maintenance of infrastructure and other assets to maximize benefits, to ensure resilience, and to manage risk in providing satisfactory levels of service to the public in a sustainable and environmentally responsible manner.

Targets:

1. Adopt a citywide asset management policy and program implementation plan by [DATE, YEAR].
2. Implement an asset management governance model by [DATE, YEAR].
3. Establish levels of service measurements consistent with asset management plans by [DATE, YEAR].
4. Develop a comprehensive 5-year capital improvement plan, and integrate resilience models by [DATE, YEAR].

Governance Theme 2: Good Governance- Accountability, Transparency, Accessibility, Community Input, and Fiscal Resilience

Goal 1:

Establish policies and tools for efficient management and effective service provision.

Outcome 1.1: Strengthen financial management processes, reporting, analysis, transparency, and control.

Targets:

1. Increase cost avoidance due to energy inefficiency by an additional [PERCENTAGE] over [FISCAL YEAR] results by [DATE, YEAR].
2. Increase the engagement of internal and external stakeholder in the diversity initiatives by [PERCENTAGE] over [BASE YEAR] by [DATE, YEAR].

Resilience Template: Emergency Preparedness and Climate Resilience

Climate change represents a set of challenges to emergency planning, readiness, and preparedness for municipalities. Communities continue to experience the effects of the extended heat events and drought. To appropriately and timely respond to heat wave events, local governments emergency action guidelines, disaster preparedness, and hazard mitigation plans are amended to include additional considerations and planning procedures incorporating the climate change information and data.

The following checklists aid the public health and communication service sections in the emergency preparedness plans:

- Coordinate with the nonprofit organizations to establish cooling shelters that provide air conditioning and relief from extreme heat events.
- Issue public statements to inform residents of ways to avoid heat-related illness and injury, encouraging them to check on vulnerable neighbors and family members and providing them with the shelter locations information.
- Issue public statement to encourage lowering emissions during prolonged heat events to improve air quality.
- Coordinate with utility companies to ensure a timely response in the case of prolonged power outages.

Beyond amendments to the local governments' emergency preparedness guidelines and plans, the list of recommendations and supporting documents are included in the Public Health Services section of the emergency and disaster plans. The items are adapted into a comprehensive report on emergency preparedness in response to extreme weather threats and the changing climate.

General Emergency Planning Recommendations:

- Inventory available means of communication and the crucial emergency personnel.
- Inventory available capacity to communicate with the community concerning emergency situations and available resources and solutions.
  - Available landlines
  - Means of contact in case of loss of landlines and cellular reception
- Assess the ability to identify specific portions of the city and population:
  - Elderly
  - Low income
  - Businesses
  - Schools
  - Community aid organizations
- Evaluate the ability to disseminate emergency information to specific parts of the population.
- Create an educational material on the likelihood of an increased number of emergency events.
- Provide an educational content on the availability of recourses for a particular emergency event:
  - Evacuation routes
  - Cooling center locations
  - Water pollution threats and potential treatments
  - Avoiding injury or health issues during high heat and poor-quality events
- Include stakeholders from various sectors to ensure all concerns are dealt with comprehensively.

Severe Heat Event Planning Recommendations:

- Use a satellite imagery, census data, and GIS to identify vulnerable population.
- Use a satellite imagery and GIS to determine locations where the heat island effect will increase severity of heat events.

- Reduce the carbon emissions during the high heat events to improve the local air quality.
- Create an automated system to alert and check on at-risk citizens.
- Open the cooling centers and collaborate with local nonprofit organizations to make temporary cooling shelters.

In the case of a protracted heat event, the flowchart represented by Fig. 7.1 features a recommended community-wide response, from coordinating with the relevant community organizations, utilities, and state, regional and federal agencies , communicating with the Public, and opening of the cooling shelters (Fig. 7.1):

Summary

Various resources and tools are available at no or at relatively small cost the local units of government to develop a robust resilience and sustainability plan. The availability and ease of access to resources make it a compelling reason for local governments to pursue resilience and sustainability planning. However, only with the supportive elected and appointed leadership and dedicated staff will those efforts to utilize available resources and tools result in concrete plans for the organization. Moreover, the available resources and tools are only relevant if the appropriate internal and external stakeholders are identified, environmental scan is conducted, and measurable targets are created using the

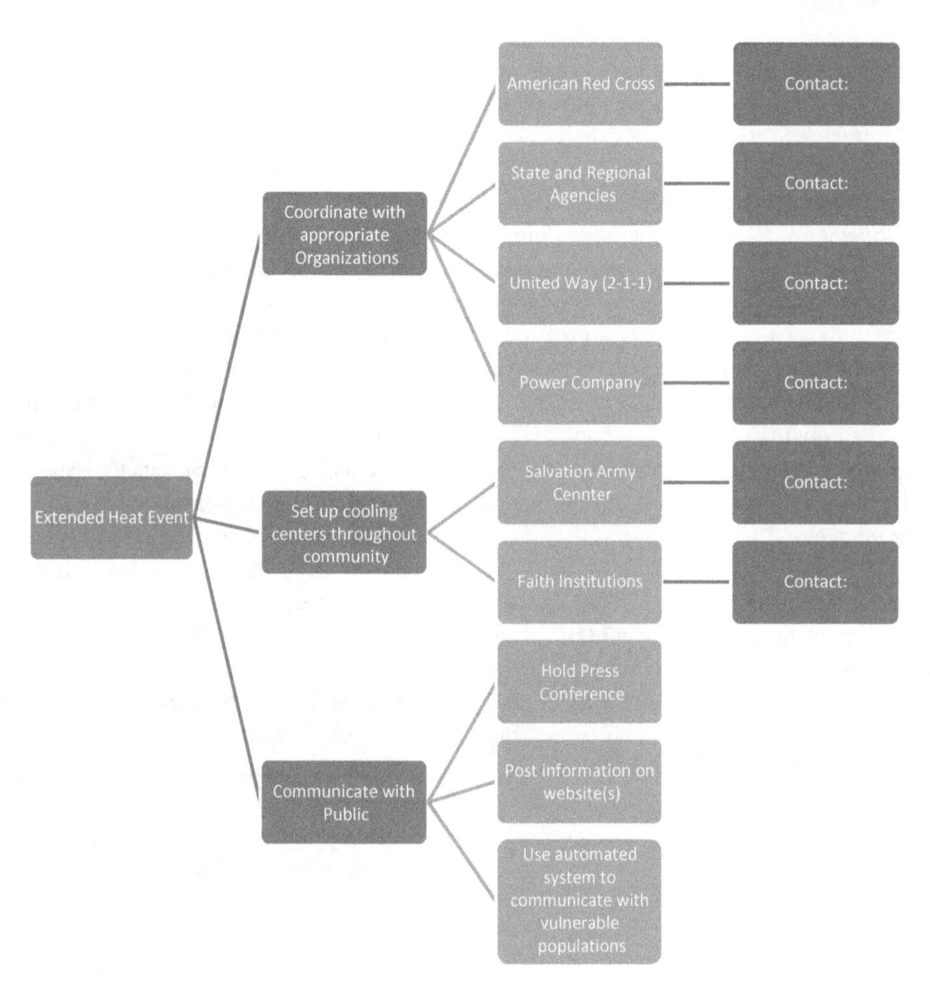

Fig. 7.1 Flowchart – emergency preparedness for extended heat wave events

steps described in the previous chapters. Finally, the tools and resources are only valuable if a local government management dedicates staff to planning and implementing a sustainability and resilience plan.

Further Discussions

- Discuss the resources that local government administrators would require for sustainability and resilience planning.
- Assess how the local governments propose to measure carbon footprint vs. energy use.
- Scrutinize the available measures utilized by local governments.
- Evaluate the available staffing and the requirements for implementing a sustainability and resilience plan.
- Analyze the objectives and targets created by a local government, including the reduction of energy consumption, renewable energy, and greenhouse gas emission targets.

References

100 Resilient Cities (100 RC) (2018) About us. Retrieved from http://www.100resilientcities.org/about-us/

Alibašić H (2017) Measuring the sustainability impact in local governments using the quadruple bottom line. Int J Sustain Policy Pract 13(3):37–45

Alibašić H (2018a) Leading climate change at the local government level. In: Farazmand A (ed) Global encyclopedia of public administration, public policy, and governance. Springer International Publishing AG, Cham. https://doi.org/10.1007/978-3-319-31816-5_3429-1

Alibašić H (2018b) Role of corporations in addressing climate change. In: Farazmand A (ed) Global encyclopedia of public administration, public policy, and governance. Springer International Publishing AG, Cham. https://doi.org/10.1007/978-3-319-31816-5_3429-1

Alibašić H (2018c) Ethics of resiliency in crisis management. In: Farazmand A (ed) Global encyclopedia of public administration, public policy, and governance. Springer International Publishing AG, Cham. https://doi.org/10.1007/978-3-319-31816-5_3426-1

Alibašić H (2018d) Ethics and sustainability in local government. In: Farazmand A (ed) Global encyclopedia of public administration, public policy, and governance. Springer International Publishing AG, Cham

Architecture 2030 (2015) The 2030 challenge for planning. Retrieved from: http://architecture2030.org/2030_challenges/2030_challenge_planning/

Architecture 2030 (2017a) 2030 Districts. Retrieved from: http://architecture2030.org/programs/2030-districts/

Architecture 2030 (2017b) About us. Retrieved from http://architecture2030.org/about/

City of Houston (2017) EPA's climate change is real website. Retrieved from http://www.greenhoustontx.gov/climate-change-is-real.html

City of Milwaukee (2017) Climate change is real. http://city.milwaukee.gov/sustainability/About-Us/Climate-Change.htm#.WmzTpZM-fys

Compact of Mayors (2018) Compact of mayors. Retrieved from http://www.c40.org/programmes/compact-of-mayors

Global Covenant of Mayors (n.d.) Welcome to the global covenant of mayors. Retrieved from http://www.globalcovenantofmayors.org/?welcome

Great Lakes Climate resilience (2013) Great Lakes climate resilience planning guide. Retrieved from http://greatlakesresilience.org/

Great Lakes Coastal Resilience (2013) Great Lakes coastal resilience planning guide. Retrieved from http://greatlakesresilience.org/

Great Lakes Saint Lawrence Cities Initiative (GLSLCI) (2016) https://glslcities.org/initiatives/climate-change-mitigation/

Ibn Khaldun AA. bin M (1406, 1969) The Muqaddimah: an introduction to history; in three volumes. (Franz Rosenthal, trans.). Princeton University Press, New Jersey. Abridged edition

ICLEI – Local Governments for Sustainability (ICLEI) (2018a) Who we are. Retrieved from http://icleiusa.org/about-us/who-we-are/

ICLEI – Local Governments for Sustainability (ICLEI) (2018b) Programs. Retrieved from http://icleiusa.org/programs/

National League of Cities (NLC) (2017) National league of cities launches local climate solutions engagement program. Retrieved from http://www.nlc.org/article/national-league-of-cities-launches-local-climate-solutions-engagement-program

National League of Cities (NLC) (2018) Environment and sustainability. Retrieved from http://www.nlc.org/topics/environment-sustainability

National Oceanic and Atmospheric Administration (NOAA) (2017) Digital coasts tools. Retrieved from https://coast.noaa.gov/digitalcoast/tools/

National Oceanic and Atmospheric Administration (NOAA) (2018) Great Lakes water level. NOAA – Great Lakes environmental research laboratory. Retrieved from https://www.glerl.noaa.gov/data/wlevels/levels.html

National Oceanic and Atmospheric Administration (NOAA) (n.d.) U.S. climate resilience toolkit. Retrieved from https://toolkit.climate.gov/

Northern Gulf of Mexico Sentinel Site Cooperative (2018) Gulf Tools for Resilience Exploration Engine (Gulf TREE). Retrieved from: http://www.gulftree.org/

Resilient Communities for American (RC4A) (n.d.) Resilient Communities for America. Retrieved from http://www.resilientamerica.org/

Resilient Michigan (2017) Planning for community resilience in Michigan: a comprehensive handbook. Retrieved from http://www.resilientmichigan.org/downloads/michigan_resiliency_handbook_web.pdf

Resilient Michigan (n.d.) About resilient Michigan. Retrieved from http://www.resilientmichigan.org/about.asp

Sengupta S, Eddy M, Buckley C, Rubin AJ (2017) As trump exits Paris agreement, other nations are defiant. Retrieved from https://www.nytimes.com/2017/06/01/world/europe/climate-paris-agreement-trump-china.html?_r=0

Telegraph Reporters (2017) Trump abandons Paris climate change: how the business world reacted. The telegraph. Retrieved from http://www.telegraph.co.uk

The Nature Conservancy (n.d.-a) Climate change in the Great Lakes Region: adapting conservation efforts for a sustainable future. Retrieved from https://www.nature.org/ourinitiatives/regions/northamerica/areas/greatlakes/explore/adapting-to-climate-change-in-the-great-lakes.xml?redirect=https-301

The Nature Conservancy (n.d.-b) Climate tools. Retrieved from https://www.nature.org/ourinitiatives/regions/northamerica/areas/greatlakes/explore/climate-change-tools.xml

The Nature Conservancy (n.d.-c) Climate wizard. Retrieved from: http://www.climatewizard.org/

The State, Local, and Tribal Leaders Task Force on Climate Preparedness and Resilience (Task Force) (2014) Recommendations to the President. Retrieved from https://obamawhitehouse.archives.gov/sites/default/files/docs/task_force_report_0.pdf

The United States Conference of Mayors (USCM) (2018) Mayors climate protection center. Retrieved from https://www.usmayors.org/mayors-climate-protection-center/

The United States Environmental Protection Agency (EPA) (2017) Pollution prevention tools and calculators. Retrieved from https://www.epa.gov/p2/pollution-prevention-tools-and-calculators

Urban Sustainability Directors Network (USDN) (2018) About USDN. Retrieved from https://www.usdn.org/public/page/5/About

8 Concluding Remarks: Future of Sustainability and Resilience Planning

"Human beings are not bound to an ecosystem or territory by instinct or physiology. Our large and complex brain enabled us to learn about and exploit our surroundings and to shape our environment so that we could live within it. Our adaptability has enabled us to survive and flourish in environments as extreme as deserts and the Arctic Tundra, as well as prairies, wetland, forests, and mountain slopes. Today that adaptability has reached astonishing levels as technological development and accelerate." Page 60, Suzuki, D. (2010) The Legacy: An elder's vision for our sustainable future

Key Questions

The eighth and the final chapter of this book aims to summarize previous sections and attempts to address the following underlying premises and topics.

- What is the future of resilience and sustainability in communities and organizations?
- Beyond climate change what challenges and barriers will cities, townships, villages, and counties face in the future?
- How can local governments maintain a sustainability and resilience planning momentum?
- How can resilience be and remain embedded within organizations?

Introduction

Sustainability and resilience planning addresses the complexities and dynamics of the interconnected systems. Local government leaders around the globe seek to provide holistic, systematic, and innovative solutions for service delivery to their residents and businesses. Cities attempt to maintain and enhance their assets, the environment and human resources. Modern administrators design municipalities and the support systems from energy to transportation to address the local communities' needs and to also enhance sustainability and resilience of those cities stemming from security threats, emergencies, and climate change. Leading complex systems requires a systematic approach to problem-solving (Alibašić 2018b).

The Current Context for Sustainability and Resilience Planning

The United States' decision to withdraw from the Paris climate accord enabled the elected and appointed officials in the US cities and some states to step up their efforts to address climate change threats. Local governments have an administrative obligation to address and prepare for the climate change-related impacts. While the numbers on world energy use and carbon emissions vary from source to source, they are staggering in size. Cities are responsible for over 75% of the world's energy use and emit more than 75–80% of all greenhouse gas emissions, mostly CO2 (Satterhwaite 2008). In meeting their goals and objectives to increase the resilience of communities, local governments face various obstacles beyond financial obligations

H. Alibašić, *Sustainability and Resilience Planning for Local Governments*, Sustainable Development Goals Series, https://doi.org/10.1007/978-3-319-72568-0_8

and challenges. Administrators make the necessary readjustment to planning of cities with an emphasis on building resilience. Most initiatives by local governments are aimed at reducing the GHG emissions, mitigating the impact of climate change and adapting to new climate realities, using cutting-edge approaches to climate resilience in cities.

By incorporating climate preparedness into their sustainability plans, strategic plans, and emergency action guidelines, cities are able to better withstand negative impacts from extreme weather events and climate change. The sustainability and resilience planning takes into consideration all the elements of the local government's systems for service provisions, including power, infrastructure, buildings, health, and safety, and incorporates them into a single resilience plan. This integrated, multi-modal, multi-system approach is described in Fig. 8.1, where the elements and processes of climate mitigation and adaptation are interspersed and connected. The Fig. 8.1 represents the Quadruple Bottom Line approach to resilience planning with the social, economic, environmental, and governance pillars successfully integrated into the strategic process. The key components include the community engagements, partnerships, reporting, and continuous improvement. Administrators within their organizational and leadership capacity have a vested interest in continuous improvement and adoption of strategies to address various stressors as they better prepare for disasters. Failure to pursue long-term resilience could have far-reaching adverse consequences on organizations and communities in case of natural disasters and crises.

The Future of Resilience and Sustainability Planning

The primary drivers of thriving sustainability and resilience programs is the ability of communities and organizations to transform and adapt to the changes in the environmental, societal, governing, and economic conditions surrounding them. In describing the culture strategy, Osborne and Plastrik (1997) posited the "organizations have distinctive personalities," adding "like people, some organization are energetic, creative, or caring, while others are depressed compliant, or neglectful," concluding that "and like personalities, organizational cultures are very hard to change" (p. 255). The cultural changes that are necessary from within and from outside are critical for successful implementation of sustainability and resilience in organizations. Both the private and the public sector organizations use sustainability to address their constituents' needs and demands. However, "government organizations are creatures of the political sector," and "inevitably are the target of incessant public demands channeled through elected officials" (Osborne and Plastrik 1997, p. 257).

Cities' leaders are engaged in sustainability and resilience to continue to thrive and provide quality of life services as revenues shrink. Local government officials are aware of the complex nature of cities and design programs to address the sustainability of organizations and to enhance resilience of communities to security threats, natural catastrophes, emergencies, and climate change. Sustainable development was defined by the Brundtland Commission, also known as the World Commission on Environment and Development (WCED) as "meeting the needs of today without comprising the ability of future generations to meet their own needs" (United Nations General Assembly 1987). In essence, sustainability may be viewed as an attempt by organizations and individuals to minimize the negative consequences of their activities on the social and environmental facets of the society while improving governance and maximizing the economic gains. At the same time, resilience may be defined as an extension and expansion of sustainability efforts in recognition of existential threats of climate change.

Sustainability and resilience planning enables organizations to use a multifaceted, cross-sectoral approach for the betterment and operational efficiency of organizations around the globe, from cities to multinational corporations. Local governments in their organizational and leadership capacity will continue to implement resilience and sustainability strategies to address environ-

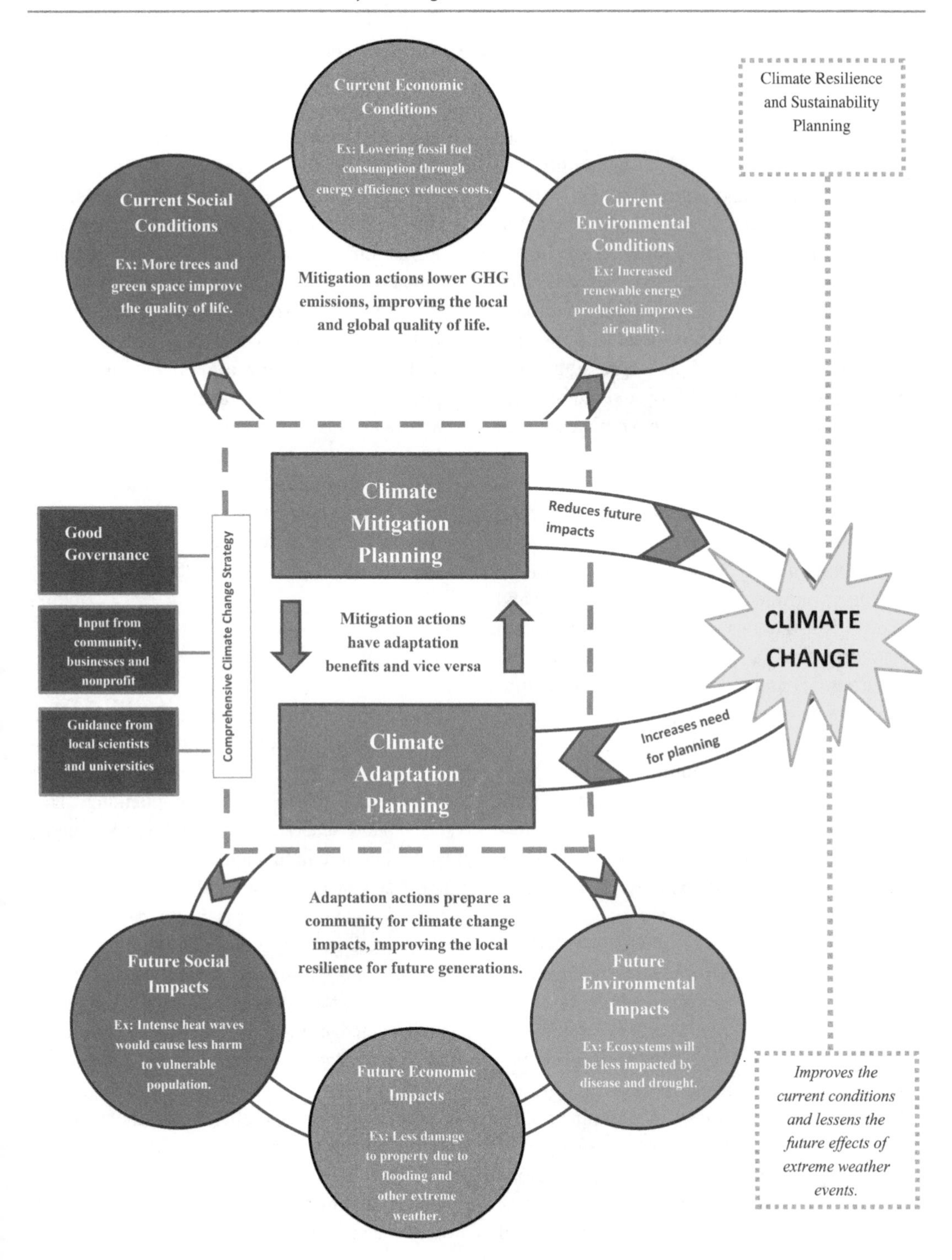

Fig. 8.1 Resilience planning through climate mitigation and adaptation strategies

mental, social, economic, and governance issues, stemming from climate change and natural disasters. Communities and organizations viewed through the system lens offer practical solutions through active utilization of sustainability and resilience programs and policies. Often, without a system-wide approach. the practical applications of sustainability and resilience are lost in the administrative and collective nuances. The sustainability and resilience planning approach yields multiple benefits to organizations and communities, summarized as addressing complexities, allowing adaptability and innovation, and encouraging good governance and responsiveness.

1. *Sustainability and Resilience Planning Addresses Complexities and the Modern Dynamics of the Interconnected Systems.* Local government leaders seek holistic and innovative solutions in an attempt to maintain and enhance the organizational assets, the environment and human resources. The complex nature of cities require the integrated approach to transportation, energy, and other elements of resilient communities to address and to enhance resilience initiatives from security threats, extreme weather, fiscal pressures, emergencies, and climate change. Leading complex systems requires a systematic approach to problem-solving.
2. *Sustainability and Resilience Planning Allows Adaptability and Innovation in Organizations.* Sustainability and resilience planning offers an opportunity for organizations to adapt to changing circumstances surrounding organizations. In its effort to provide innovative solutions, local community administrators integrate a holistic approach to service delivery. Furthermore, "modern organizations design their systems using effective sustainability strategies to withstand external and internal pressures for maximum resiliency in dynamic environments, as defined by the threats of climate change, global warming, and growing existential economic, environmental, governance, and societal pressures" (Alibašić 2018a , p. 1).
3. *Sustainability and Resilience Planning Encourages Good Governance and Responsiveness.* Changes in the external and internal organizational factors and societal priorities lead to increased responsiveness, identified priorities, and shared responsibilities, adjusted approaches undertaken by leadership in responding to demands and engaging on issues. Local governments invest in long-term sustainable and resilient initiatives to improve overall effectiveness and efficiency of the organizational service delivery with a solid return on investment. As an illustration, cities invested heavily in sustainable energy. Besides, "sustainable energy policies, programs, and projects at the local and state levels of government represents an inevitable challenge, but also a conceivable opportunity" (Alibašić 2017b, p. 1).

Summary

The research presented in this book goes beyond theoretical and practical framework and postulates to provide an overview for thriving and lasting sustainability and resilience planning. The research builds and relies on a vast body of already existing literature in the field of sustainability and climate resilience, with specific functional design features for elements necessary for development and application of sustainability and resilience initiatives and programs. Additional, in-depth review of existing programs, resilience plans, sustainability and climate action efforts was provided. The previous seven chapters provided the detailed outline to initiating, developing, and implementing a sustainability and resilience plan, with the following elements outlined:

- Understanding the differences and similarities between sustainability and resilience
- Initial mapping out the sustainability and resilience processes for organizations and communities
- Identifying the internal and external stakeholders, the level of organizational and community engagement, and outcome champions

- Measuring, tracking, monitoring, and reporting sustainability and resilience progress using the Quadruple Bottom Line (QBL) mechanism
- Implementing the sustainability and resilience plan, illuminating programs and initiatives to achieve sustainable and resilient communities
- Examining the intersection of sustainability and resilience and the corresponding outcomes for diverse planning processes
- Exemplifying innumerable tools and resources for sustainability and resilience planning

While resilience planning represents the next stage in sustainability planning and includes the elements of sustainability, understandably some organizations may continue to utilize sustainability plans for the consistency and continuity sake. However, in such instances, it is imperative the sustainability plan be reinforced with climate adaptation, climate mitigation strategies, and climate preparedness actions to achieve resilience.

The path to sustainability and resilience for local government requires a confident interplay of organizational dynamics of leadership and financial commitment, and implementation of policies and programs. The following non-exhaustive list of dynamic, sustainability and resilience components required for sustainable and resilience programs, policies, and projects includes:

Commitment Organizations are committed financially and otherwise to the sustainability and resilience goals and practical application and implementation of resilience and sustainability-related efforts. By connecting sustainability and resilience planning directly to their budget process, organizations indicate and substantiate commitment to sustainability and resilience.

Leadership Just as with financial responsibility, the key to successful sustainability and resilience program is capable and willing leadership. There is a direct correlation between exemplary leadership, sustainability, resilience, and successful organizations. Elected and appointed officials and staff must work in sync. Leaders enable and encourage development and implementation of sustainability and resilience initiatives at all levels of organizations. Evaluation of the leadership role is provided within the framework of a broader governing perspective. By incorporating sustainability and resilience measures into its core value, an organization gets a more defined focus on long-term planning and leadership development.

Measurement, Monitoring, and Tracking Measuring sustainability and resilience impact is a necessity. Administrators view the the resilience outcomes in the context of budgetary implications are and track and measure results in real time. Measurement allows for better accountability and transparency of sustainability and resilience programs. The design of the sustainability and resilience program does not necessarily lead to more efficiency unless it is designed to be tracked, measured, monitored, and reported (Alibašić 2017a).

Reporting Mechanism Final and interim results are published and incorporated into future budgetary priorities and fiscal plans. By presenting outcomes from sustainability and resilience programs, organizations exercise a democratic right to communicate to constituents what they stand for, what needs to be improved and the plans for the future.

Stakeholder Engagement The stakeholders, both internally and externally, are more inclined to support and be engaged with the sustainability and resilience initiatives and programs, when they understand how sustainability enhances service delivery, and how it improves organizational efficiency and the quality of life in communities. The key to such engagement is to identify the uniqueness of the local government organization and the community.

Partnership Prosperous programs and initiatives involve innovative and pragmatic alliances to offset and leverage resources in the world fraught with financial instabilities. Financing resilience and sustainability-related projects requires the engagement of the banking sector, private sector, and public sectors in ways not seen in the past. One of the fundamental characteristics in the adoption of sustainability is the jump from planning to implementation. The partnership between local governments, academic institutions, and private sectors can facilitate such jumpstart.

The process of embedding sustainability and resilience at all levels of local government is arduous; requires resources, commitment, and adaptation; and is not done in a vacuum. Each segment of creating sustainable and resilient community is critical. Ethical implications of planning and implementing resilience and sustainability-related initiatives using the Quadruple Bottom Line approach includes more just, equitable, healthy, and economically and environmentally resilient communities with an overall positive societal outcome.

Further Discussions

- Assess the implementation strategies for sustainability and resilience projects.
- Analyze the impact of sustainable energy strategies in organizations.
- Discuss the types of sustainability and resilience projects.
- Evaluate the implementation strategies and QBL approach to implementation.
- Examine the future of resilience planning and the Quadruple Bottom Line approach.

References

Alibašić H (2017a) Measuring the sustainability impact in local governments using the Quadruple Bottom Line. Int J Sustain Policy Prac 13(3):37–45

Alibašić H (2017b) Sustainable energy policy for local and state governments. In: Farazmand A (ed) Global encyclopedia of public administration, public policy, and governance. Springer International Publishing AG, Cham

Alibašić H (2018a) Redesigning organizations for maximum resiliency in dynamic environments. In: Farazmand A (ed) Global encyclopedia of public administration, public policy, and governance. Springer International Publishing AG, Cham. https://doi.org/10.1007/978-3-319-31816-5

Alibašić H (2018b) Sustainability as organizational strategy. In: Farazmand A (ed) Global encyclopedia of public administration, public policy, and governance. Springer International Publishing AG, Cham. https://doi.org/10.1007/978-3-319-31816-5_3433-1

Osborne D, Plastrik P (1997) Banishing bureaucracy: the five strategies for reinventing government. Addison-Wesley Publishing Company, Reading

United Nations General Assembly (UNGA) (1987) Report of the world commission on environment and development: our common future. Transmitted to the General Assembly as an Annex to document A/42/427 - Development and International Co-operation: Environment. Retrieved from: http://www.un-documents.net/ocf-02.htm

Satterhwaite D (2008) Cities' contribution to global warming: notes on the allocation of greenhouse gas emissions. Environ Urban Int Inst Environ Dev (IIED) 20(2):539–549. https://doi.org/10.1177/0956247808096127

Suzuki D (2010) The legacy: an Elder's vision for our sustainable future. Greystone Books D&M Publishers, Inc, Vancouver

Index

H. Alibašić, *Sustainability and Resilience Planning for Local Governments*, Sustainable Development Goals Series, https://doi.org/10.1007/978-3-319-72568-0

N

O

P

Q

R

S

CPSIA information can be obtained
at www.ICGtesting.com
Printed in the USA
LVHW06*1333200818
587517LV00004B/9/P